Collins

How to be a
SOCIAL
RESEARCHER
Using Sociological Studies

Sarah Cant and Jennifer Hardes Dvorak

William Collins' dream of knowledge for all began with the publication of his first book in 1819.

A self-educated mill worker, he not only enriched millions of lives, but also founded a flourishing publishing house. Today, staying true to this spirit, Collins books are packed with inspiration, innovation and practical expertise. They place you at the centre of a world of possibility and give you exactly what you need to explore it.

Collins. Freedom to teach.

Published by Collins

An imprint of HarperCollins*Publishers*
The News Building, 1 London Bridge Street, London, SE1 9GF, UK
Macken House, 39/40 Mayor Street Upper, Dublin 1, D01 C9W8, Ireland

Browse the complete Collins catalogue at
collins.co.uk

© HarperCollins*Publishers* Limited 2023

10 9 8 7 6 5 4 3 2 1

ISBN 978-0-00-855468-2

All rights reserved. No part of this publication may be reproduced, stored in a retrieval system, or transmitted in any form by any means, electronic, mechanical, photocopying, recording or otherwise, without the prior written permission of the Publisher or a licence permitting restricted copying in the United Kingdom issued by the Copyright Licensing Agency Ltd, 5th Floor, Shackleton House, 4 Battle Bridge Lane, London SE1 2HX.

British Library Cataloguing-in-Publication Data

A catalogue record for this publication is available from the British Library.

Dedication: For George

Authors: Sarah Cant and Jennifer Hardes Dvorak
Publisher: Katie Sergeant
Product manager: Catherine Martin
Development editor and copyeditor: Sonya Newland
Proofreader: Claire Throp
Permissions researcher: Rachel Thorne
Cover designer: Ken Vail Graphic Design
Cover image: Laura Buckley, *Fata Morgana*, 2012 © Laura Buckley, photo: emmak95/Stockimo/Alamy Stock Photo
Internal design and illustration: Ken Vail Graphic Design
Production controller: Alhady Ali
Printed and bound by Replika Press Pvt. Ltd.

This book is produced from independently certified FSC™ paper to ensure responsible forest management.

For more information visit: www.harpercollins.co.uk/green

Acknowledgements

With thanks to the following students, teachers and academics for reviewing sections of the book: Dr Rima Saini, Senior Lecturer in Sociology, Middlesex University London; Balkar Gill, Head of Sociology, Joseph Chamberlain Sixth Form College, Birmingham; Annalisea Arthur-Whyte, Associate Assistant Headteacher, London; Lesley Connor, sociology teacher, The Sixth Form College, Farnborough; Zach Dunscombe (Year 12 student), Freya Talbot (Year 12 student) and Fran Nantongwe, sociology teacher at Notre Dame High School, Norwich.

CONTENTS

Introduction	What is social research?	4
Chapter 1	Becoming a social researcher	10
Chapter 2	Doing your own social research	33
Chapter 3	Official statistics	42
Chapter 4	Social surveys	66
Chapter 5	Interviews	96
Chapter 6	Observations	121
Chapter 7	Documentary methods	146
Chapter 8	Experiments	166
Chapter 9	Mixed methods: Society as a kaleidoscope	185
Chapter 10	Shaking up methods	201
Chapter 11	A final turn of the kaleidoscope: The enduring importance of social research	217
References		221
Glossary		226
Index		231
Acknowledgements		239

INTRODUCTION: WHAT IS SOCIAL RESEARCH?

Our human-made world is complex, changing, often surprising and endlessly fascinating. It is the job of the sociologist not only to make sense of the kaleidoscopic social world, but also to imagine better, alternative futures. This gives sociological research both focus and ethical responsibility.

But what do we mean by the 'social world'? And how can we study it? 'Social' includes everything that is organised, structured and experienced by humans – anything and everything that people do. Sociologists focus on human social interactions – the ways in which people live and communicate in families, groups, institutions, communities and societies. But, more than this, sociologists are interested in making connections between these myriad social interactions and the wider world. They examine how processes of social change, the historical context, economic systems, cultural values, social relationships and networks, political ideologies, technological developments, the natural environment and even our relationships with animals all impact on human activity, shape life chances and explain diverse patterns of inequality.

Sociology, then, is the study of the relationship between individual experiences and wider social structures – between self and society, between biography and history – and sociologists need research methods to help them understand this intimate and intriguing connection. For example, during the COVID-19 pandemic, many people experienced loneliness, sickness, job redundancy, bereavement and anxiety. Sociology tries to understand how people navigated and managed these experiences, but also how the experiences were linked to economic, political, cultural and social factors. Sociologists used research methods to investigate how social class, gender and ethnicity predicted COVID-19 outcomes. They also sought to understand and map how broader economic policies and geographical locations shaped people's access to social and medical support (such as vaccinations) during the pandemic.

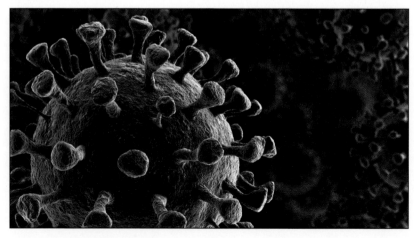

Of course, sociologists recognise that social research is not limited to the investigation of personal troubles. It also examines people's personal truths and triumphs – their beliefs and their successes. For example, sociologists examined why some people were afraid of being vaccinated and why others viewed COVID-19 as a conspiracy. They also explored who was able to easily work from home with the advantage of outdoor space, additional family time, and the opportunity to save money.

Case study: Sociology of the pandemic

Sociologists have used a range of research methods to investigate the impact of the pandemic. For example, Matthewman and Huppatz (2020) explored the insights that a sociology of COVID-19 offered, describing the pandemic as a giant sociological experiment that ushered in dramatic changes to social life. Communities came together, helping one another, while social inequalities widened. New rules for social conduct were implemented and opportunities for social interaction were curtailed. The health services and the economy were put under considerable strain. Berghs (2021) asked 'Who gets cured?', pointing to inequalities in infection, prevention and intervention, especially since access to ventilators and the roll-out of the vaccines were far from even, both around the globe and within countries. Maestripieri (2021) examined COVID-19 inequalities, showing how migrants and minorities, women, children, disabled people and those in lower social classes were disproportionately affected. In this way, the pandemic was not a leveller but a divider, exposing the fractured nature of human society (Scambler, 2020). Inequalities stretched into the home, as gender-based violence increased across the globe (Dlamini, 2021) and, as schools closed their doors to most students, the burden of childcare also mainly fell on women (Yavorsky et al. 2021).

Sociologist Melinda Mills undertook several important studies. With colleagues Ding and Brazel, her survey data (2022) showed how COVID-19 affected everyone's sleep but with a bigger impact on women. Although men experienced some sleep loss because of their changing employment and financial status, women's sleep was impacted more extensively by increased levels of anxiety and the pressures of home-schooling and caregiving. In a research team led by Jennings (2021), Mills was also involved in five focus groups to understand why some people in the UK were hesitant about having the vaccine. The research showed that those people who mistrusted the government and relied on information through unregulated social media were more reluctant to be vaccinated. These results were fed back to the government along with suggestions that official advice about vaccination should also be circulated through social media.

This book explores the range of research methods that sociologists use to understand how society shapes personal troubles, truths and triumphs, using both classic and contemporary research studies to show how the methods can be applied. Examining the actual 'doing' of research reveals the joy, value and messiness of sociological work, the ethical dilemmas of practical study, and the checks that must be put in place to protect research subjects and guard against the researcher's own **biases**. The chapters that follow highlight the advantages and disadvantages of different methods, the insights that can be gained, and the impact that sociological research has on those being studied as well as on wider society. By examining sociological studies you will engage with the practicalities of research, which will help you to become a reflexive and careful researcher. It will also enhance your knowledge of key areas of sociological enquiry and important sociological concepts, to support your studies at school, college and university.

Why does sociology need its own research methods?

All academic subjects have different ways of gathering data, information and evidence. Historians might interrogate documents from the past or listen to people's memories. A scholar of English literature may set a text in its social and historical context. Astrophysicists study space with telescopes, but they also hypothetically model what they think might be happening in black holes, as they cannot see inside them. A geneticist may use microscopes to profile chromosomes, genes and DNA, looking for patterns, enabling them to model risks and to develop medical interventions. In the same way, sociologists have a set of methods – their very own toolbox – that they use to study society.

Studying research methods is both important and fascinating. Without a toolkit of research methods, how would we know about the inequalities revealed through the pandemic, different outcomes in education or the unequal division of domestic labour? How would we know whether crime rates are going up or coming down, or who is more likely to be a perpetrator or a victim of crime? How could we be sure whether income for the same job varies by gender, ethnicity and age, and in what ways? How could we judge whether we live in a tolerant or a prejudiced society? How would we know why people vote in the way that they do, why divorce rates are increasing and why people are having fewer children?

Sociology is the study of both the large-scale social forces that shape our lives and the small-scale social interactions that give our lives meaning. Therefore, researchers need ways of gathering evidence and data that can help them to understand both **macro** and **micro** aspects of human society. They need methods that can tell them about how society is organised and divided, and the impact that often-hidden social structures have on people's life chances, such as the ways that education systems, laws, **capitalism**, **patriarchy** or **institutional racism** influence success in education and work. Sociologists also need methods that can draw out people's experiences of living in the world – their stories, beliefs, values and understandings.

Approaches to sociological research

In simple terms, social researchers make a distinction between two approaches to studying the social world – **quantitative** and **qualitative**. Quantitative research aims to assert facts about the social world using methods that enable the researcher to gather numerical data (such as surveys). Sociologists can then analyse this data using statistics. Quantitative research typically focuses on large-scale (macro) social patterns, sometimes making generalisations about the world we live in. By contrast, qualitative research aims to develop a more detailed, interpersonal understanding of the social world. It uses methods such as interviews and **ethnographies** to elicit and make sense of micro-level interactions between people.

Sociologists are are sometimes divided over which research approach to take because they recognise that the distinctions between quantitative and qualitative approaches are not clear-cut. Quantitative researchers usually focus on revealing facts and patterns. Qualitative researchers tend to believe that social life is too messy and complex to be able to identify generalisable patterns. Instead, they try to reveal more about people's beliefs, experiences and understanding of their everyday social lives. Sociologists decide which approach to take based on the theoretical perspective they adopt – these are explored further in Chapter 1.

> **Key concepts**
> **quantitative methods** methods of research that produce numerical data
> **qualitative methods** methods of research that produce written data
> **ethnography** when a researcher immerses themselves in a field to gain an understanding of life in a 'naturalistic' setting; in more general terms, ethnography is the study of people and their cultures

All social research, both qualitative and quantitative, should strive to be rigorous and careful, but also aim to be impactful and meaningful. The rigour and care of research methods is what separates social researchers from other people who comment on the social world. Defending the use of research methods is essential for sociologists because they study the very world that non-sociologists inhabit. Unlike most other scientists, sociologists do not have special instruments, such as telescopes or microscopes, so they must show that what they know is not simply common sense, but is informed by careful research and diligent analysis, and thus distinct from what parents, friends, neighbours, or indeed politicians and journalists, might think about the state of the world.

Social research does not simply study and reflect on the social world; it also strives to make a real difference to people's lives. Just as geneticists develop medical interventions, sociologists suggest social interventions that can make the world a better place.

To help you judge the rigour and care of the social research that you study – and to help you plan your own research – you can use a set of research rules based on the acronym STRIVE. All good researchers should strive to conduct research that is:

S = sociological

T = theoretical

R = representative

I = impactful (transformative – that is, it makes a difference to society)

V = valid and reliable

E = ethical and reflexive

This is covered in more detail in Chapter 1.

How is the book organised?

The following chapters each explore a major sociological method, providing an overview of how it is used. Each method is illustrated using classic and contemporary sociological studies, covering research into crime, family life, education and inequality. To give you a flavour of both the capacity and limits of each research method, we use the repeated example of studying housework. Each chapter also includes researcher insights and activities to get you thinking in more detail about each method and to prepare you for studying at university. Each chapter concludes with a series of summary questions and a synopsis of the advantages and disadvantages of the method under discussion.

Chapter 1 examines the STRIVE rules of research methods in more depth, exploring the relationship between theory, methods and knowledge, and reflecting on whether sociology can be considered a science. This chapter can be read as a standalone and then revisited once you have engaged with the remainder of the book.

Chapter 2 outlines the steps required to undertake your own research, from carrying out literature reviews, devising research questions, operationalising concepts, administering pilots and collecting data, to analysing and writing up your findings.

Chapter 3 looks at one source of **secondary data** – official government statistics. For example, statistics from the Census, the British Crime Survey or the British Social Attitudes Survey can be used uncritically as data or interrogated to assess their **reliability** and **validity**. The chapter starts with Durkheim's famous study of suicide. It then explores two contemporary research studies that have shown how people of working-class origin are held back by a 'class ceiling' while middle-class people are protected by a 'glass floor'. Finally, studies that have examined the police practice of stop and search and domestic violence statistics help us to understand the criminal justice system and familial power relations.

Chapter 4 looks at another (largely) quantitative method – the use of surveys and questionnaires. Those that are developed and administered by researchers themselves are examples of **primary data**. If surveys use **closed questions** they are examples of quantitative research. However, with the inclusion of **open questions**, surveys can provide more detailed understanding – a quasi-qualitative method. This chapter looks at classic studies by Du Bois, Bourdieu and Townsend. It then explores more contemporary research, including the Great British Class Survey, the impact of putting children into sets at school, the experience of being pulled over for car offences, and male infertility.

Chapter 5 turns to interviews as a research method. Although interviews can be structured, producing quantitative data, they are usually semi-structured or unstructured. Unstructured interviews are designed to elicit detailed meanings and experiences from the perspective of the interviewees. This style of research can be particularly useful for exploring under-researched or sensitive areas. Oakley's research that interviewed women about housework and motherhood provides the classic research studies for this chapter. The contemporary studies focus on research into doorstep crime, the educational experiences of British Muslim girls and boys, women who choose to give up work on becoming a mother, and parenting in British Chinese families.

Key concepts

secondary data data that already exists, typically in the public domain (e.g. official statistics)

reliability the repeatability of research, which is considered reliable if different researchers could achieve the same result by using the same method

validity the extent to which the research measures what it intends to measure; that is, do the findings represent social reality?

primary data new data that is created by a researcher through the use of research methods, e.g. data produced through an interview or survey

closed questions questions that give respondents limited choices to select from, collecting quantitative data

open questions questions that require a participant to elaborate and answer in their own words, producing qualitative data

Chapter 6 focuses on observational research, covering several distinct types of research practice. Observations can be quantitative (where the researcher might count the number of times something happens) and qualitative (an observation of all activity with an open mind). The latter can involve non-participant and participant observation and can be overt as well as covert. Qualitative observations are sometimes referred to as an ethnography. This chapter introduces Humphrey's classic 'tearoom trade' study, and then details contemporary ethnographies of gang culture, arranged marriage, educational progress, and the use of food banks.

Chapter 7 looks in detail at documentary methods, which analyse written or spoken language in an archive, such as a set of policy reports, newspapers, magazines or diaries. While a researcher can use quantitative methods to undertake a content analysis – for example by noting how many times a particular word is used – researchers using this method can also undertake **discourse analysis**, in which they look for themes and shifts in language and meaning. For example, a researcher might examine the cultural assumptions revealed in texts and images. This chapter also outlines how sociologists use social media and **big data** as an archive. Certainly, the growth of social media has been so dramatic and ubiquitous that sociologists cannot ignore it as a source of data. The chapter introduces Foucault's approach to discourse analysis and Hall's classic research into how mugging was portrayed in the media, creating a **moral panic**. More contemporary studies analyse pregnancy apps, the portrayal of female murders in the media, educational policy change, the depiction of people living in poverty in newspapers, and girl guiding.

Chapter 8 outlines the experimental method in sociology, showing its enduring, although less widespread, application. Garfinkel's breaching experiments and Elliott's exploration of classroom prejudice are given as examples of classic research. The chapter also reviews contemporary experiments that have examined bias in classrooms, assessed parental attitudes to gender nonconformity, societal views of atypical parents, and tested 'broken windows' theories of crime.

Chapter 9 reflects on social research methods and shows why many sociologists choose to use **mixed methods** in practice. The chapter looks at Barker's classic study of a religious sect, supplemented with contemporary examination of inequalities in family sleep patterns, what it means to be labelled a 'boffin' in school, and drink-spiking.

Chapter 10 explores cutting-edge research methodology that draws on **autoethnographic** and **ethnobiographic** methods, research that incorporates 'fleshy' (corporeal) **embodied** methods, post-human approaches, and methods that highlight the importance of **decolonial methods** and **queer theory** perspectives.

Finally, **Chapter 11** makes a case for the enduring importance of sociological enquiry. With such widespread access to data now available, having a sociological imagination, sociological interpretative skills, and sociological care and reflexive consideration has never been more important.

Throughout the book, we hope you will see how learning about research methods is interesting and illuminating for your study of sociology. More than this, though, we hope that it provides you with the tools to become a rigorous, careful and ethical social researcher yourself.

Key concepts

mixed methods using more than one method to study a social phenomenon

autoethnography a method of study that foregrounds the personal experience and autobiographical accounts of the researcher, with ethnographic approaches to research that study, describe and theorise the wider cultural context

ethnobiography a method of study that foregrounds the experience of one other person (biography) but that also draws on ethnographic research (observations, interviews) to gather data and situate this biographical account in a wider social context

decolonial methods research methods that recognise that research is often conducted from western/global northern perspectives, which tend to ignore or subordinate Indigenous/Global Southern knowledges and fail to question dominant power relations

CHAPTER 1: BECOMING A SOCIAL RESEARCHER

Society is the subject matter of sociology, but the social world is notoriously difficult to study. Society is constantly changing, in response to events such as war, economic crises, new technologies, environmental challenges or a global pandemic, for example. In turn, humans constantly adapt how they live together, they change the ways that they work and organise their leisure time, and even alter how they manage birth and death.

While this makes sociology an endlessly fascinating subject to study, it also means that no sociologist can ever hope to understand the full enormity of social life. Therefore, sociologists need many different ways of gathering, analysing and interpreting data. For example, some sociological questions might relate to how policies are formed and how they operate, while others might focus on the everyday experiences of ordinary folk. These questions require different research methods. Studying a political party's commitment to recycling would require a different approach to data collection and analysis than research into the extent to which households engage in recycling. The constant changes within society also mean that new methods are sometimes needed to study it. Much social interaction, for instance, is now conducted via social media on smartphones, so new sociological research techniques are required to analyse these interactions.

Even when sociologists focus on a single issue, they often disagree about what is going on – they are guided by different theoretical perspectives and ask different questions. Famously, Durkheim (1897/2005), a **positivist** researcher, focused on the social causes of suicide. In contrast, **interpretivist** sociologists have been more interested in how deaths become classified as a suicide in the first place.

This is an important chapter, which sets out the rules and requirements for social research. It can be read on its own and then revisited to help clarify ideas and concepts.

What influences social researchers?

It is important to recognise that sociologists' research interests and methods are shaped by time and place. For instance, in the nineteenth century, sociologists typically studied men's lives – a reflection of the entrenched gender inequality that characterised that century and the discipline of sociology itself. Feminists drew attention to this omission, and developed a range of **feminist perspectives** and **feminist methods** to capture the experiences of women. Even today, sociologists remain blinkered to certain issues and questions. For example, they are being challenged to recognise how their ideas and questions are still shaped by the enduring impact of **colonialism** and colonial ways of thinking. With this observation comes the responsibility to develop and use decolonial methods. Sociologists also recognise that their thinking and research has largely been shaped by prevailing norms about sex, gender and sexuality. In response, they

---

Key concepts

positivism a perspective that assumes sociology can and must emulate the natural sciences such as physics, chemistry and biology; the notion that we can understand the social world objectively, free from any subjective bias or interpretation of events

interpretivism a perspective that suggests we can only gain knowledge about the social world through interpretations of observable phenomena

feminist methods research methods that seek to explore the experience of women and place women at the centre of research

have developed **queer methods**, underpinned by **queer theory**, to challenge the dominance of **heteronormativity** and **cisnormativity** in research.

As you can see, the social context in which sociologists live and work shapes the methods of research at their disposal, and frames their decisions about which ones to use. This insight requires that sociologists acknowledge how their social position (their class, gender, ethnicity, sexuality, etc.) influences their dispositions, the questions they ask, the methods they use and the conclusions they reach. Sociologists refer to this as research **positionality**. Their questions and methods are also sometimes shaped by who funds their research.

Sociologists also must assess the usefulness and relevance of the theories and concepts that they draw upon – this is referred to as being **reflexive**. Sociologists acknowledge that sometimes concepts can become redundant, out-dated or even dangerous (Cant and Chatterjee, 2022). Beck (2002) coined the term **zombie categories** to refer to concepts that are no longer applicable, are too broad to be meaningful, or which carry assumptions that can be challenged ('dead'), but that sociologists sometimes keep 'alive', like zombies. Beck used the example of social class to illustrate this concept.

Overall, a researcher's positionality and their theoretical perspective (for example, feminism) influences their research decisions. This means that sociologists must explore **methodology** (the theory of methods) to reveal the connections that exist between knowledge, position, power and method. As such, this chapter focuses not on the methods themselves, but on what guides a researcher's choice of methods. This understanding is key to becoming a good social researcher yourself.

Key concepts

queer methods research methods that draw from queer theory and aim to challenge existing (and often binary) categories and concepts, particularly in relation to gender, sex, and sexuality, that are instead seen as fluid, negotiated and performed

positionality a person's social location and identity, defined by elements such as social class, gender, ethnicity, sexuality and (dis)abilities

reflexive (reflexivity) describing an approach in which sociologists continually subject their work (and that of others) to critical reflection

methodology the study of methods and the theoretical assumptions that underpin them

The social kaleidoscope

Think of society as a kaleidoscope – a complex, ever-changing set of patterns and relationships. Applying a different theory to society is like making a turn of the lens in the kaleidoscope. Each theory provides a different view of the patterns, colours and contours of the social world. Theory helps sociologists to impose some order on the complexity. It provides a road map of sorts – a way of making connections and seeing the bigger picture, even though some questions may remain unanswered or perhaps may not even be asked. In turn, a sociologist's disposition towards theorising society shapes the research methods they choose and, in turn, how they see and understand the kaleidoscopic patterns of social life.

These observations give rise to two important points that social researchers need to be aware of. First, social research aims to be scientific and objective. However, second, it is shaped by the social world in which it is conducted. Therefore, it is always important to reflect on how the knowledge produced by social research is insightful but also partial.

STRIVE

To help you judge social research, you can use a list of criteria that can be remembered using the acronym STRIVE.

- **S**ociological: Does the research fulfil the promise of sociological enquiry, making connections between the personal and social?
- **T**heoretical: Which perspectives are guiding the research?
- **R**epresentative: Has the researcher included enough people in their research to be confident that the results represent the experiences of all people (in other words, are the results generalisable)?
- **I**mpactful: Will the findings make a difference and improve social life?
- **V**alid and reliable: Does the research measure what it was intended to (valid), and is it repeatable (reliable)?
- **E**thical and reflexive: Is the research careful? Has the role (position) of the researcher been considered? Are all research subjects protected? Have researchers acknowledged how their own bias and blinkers might have prevented them from asking certain questions?

We are now going to explore each of these criteria in turn.

1. STRIVE: Sociological

The first rule for social research is that it must be properly sociological. That might sound obvious, but it is an important point to make because although many people engage in social research, *sociological* research is distinctive and incredibly important, despite some challenges to its authority, access and status.

Savage and Burrows (2007) argue that sociologists historically pioneered two key methods: in-depth interviews and surveys. However, with the rise of journalistic storytelling and hard-hitting documentary-making, alongside huge amounts of data (datasets) gathered by governments and organisations such as Facebook, one might ask whether we need sociologists any more. What is the role of the sociologist in social research?

There are four key elements of sociological research that make it distinctive and important.

A   Sociologists offer a unique perspective on the social world.

Sociologists ask different questions and have different objectives to psychologists and biologists, for example. Rather than rooting social problems in an individual's psyche or brain, or in their genes, sociologists turn to social factors to explain behaviour. Mills (1959/2000) describes the essence of sociology as the study of the relationship between individuals and society, which suggests a deep link between personal troubles, truths and triumphs and broader social structures. *No other discipline does this.*

💭 Thinking further

Non-sociological research might look at social issues, such as gender differences in wages or differences in educational attainment by ethnicity, but sociological research will also seek to explain these differences in relation to social causes and correlations. Thus, sociological research does much more than simply identify and map social phenomena; it also conceptualises, predicts and explains why social phenomena take particular forms, allowing sociologists to consider how social life might be changed and improved.

B Sociologists hold 'pseudo-sociology' to account.

'Pseudo-sociology' refers to all forms of enquiry that examine sociological questions but do not use robust methods. Following the observation that sociology is distinct, sociology has an important role in questioning and holding pseudo-sociology to account. Consider the British film *Sorry We Missed You* – an insightful and emotive commentary about the difficulties of being in precarious employment as a courier – and Michael Moore's film *Sicko*, on the state of health care. The content of dramas and film documentaries like these and newspaper articles may speak to sociological concerns, but it is not supported by data that has been collected according to the rules of rigour, evidence and **ethics** that sociological research demands. While journalists often use interviews to learn about current issues, the information they gather is often not rigorously researched; their priority is to make their output 'newsworthy'. Media output may be biased rather than value-free and reflexive.

> **Key concepts**
> **ethics** a set of moral principles that govern the conduct of research, such as maintaining anonymity

Case study: Is all social commentary sociological?

Television celebrity Stacey Dooley presents herself as a feminist committed to current affairs in her documentary series *Stacey Dooley Investigates*. The programme has seen her in a range of different situations covering many different topics, including Islamists, suicide bombers, a psychiatric ward, the strict lifestyle that nuns commit to and an exploration of how tourism in Thailand and Kenya affects employees. It is easy to see why Dooley's series appeals to sociology students, as it explores issues of social justice such as mental illness and medical control, the power of religion, globalisation and gender difference. Dooley's work is undoubtedly interesting, but is it sociological? This book will help you to answer this question.

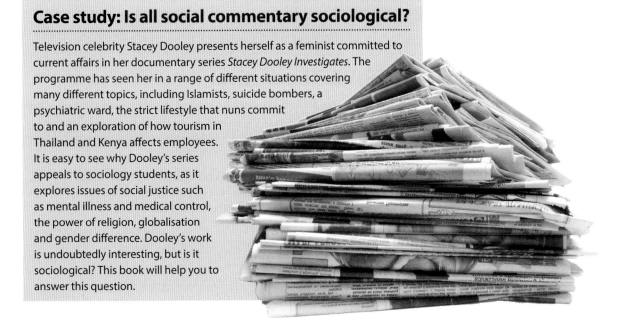

C Sociologists gather important data that is not readily available and might otherwise never be collected.

Sociologists often ask questions that other people would not think to ask, or which may have no easy answers. They unsettle accepted ways of thinking about the world and, in doing so, challenge the power of people, authorities and governments. In this way, sociologists identify the ways that power operates and give voice to groups and individuals who do not have political power in society.

D Sociologists draw out and interpret the significance of existing datasets, and challenge their inferences.

One of the most interesting recent developments has been the rise of large datasets generated by centralised surveys from governments and other official bodies. Because of this, fewer sociologists are undertaking their own surveys and are instead using secondary data and official statistics. Savage and Burrows (2007) argue that the increase in big data, especially from social media, has potentially outpaced sociologists. Sociologists have to secure funding, and they have smaller budgets. They also work on a much smaller scale and at a slower pace than the huge organisations that capture these significantly larger datasets.

However, the role of the sociologist has never been more important. Indeed, the existence of large datasets makes it imperative that sociologists apply their rigorous research and analysis skills to interpreting the data and assessing its validity. Sociologists may rely on official government data as evidence, but they also question this data, bringing a different perspective and revealing factors that may have been overlooked or under-researched. For instance, at face value, crime rates suggest that people in the working classes are more likely to commit crime than people in other social classes. However, sociological research has shown that middle-class crimes are less likely to be investigated and recorded, which brings the crime-rate data into question.

2. STRIVE: Theoretical

A sociologist's choice of method cannot be reduced to a simple weighing up of the advantages and disadvantages of the technique. It will always be guided by the questions that the researcher wants to ask and the theoretical disposition that they adopt. It is important to understand this connection.

All research is underpinned by two key philosophical concepts: **ontology** and **epistemology**. These concepts establish the kinds of questions the researcher will ask about the social world, the types of data they will gather, and how they will analyse and interpret that data.

Ontology

Ontology means 'theory of being' or 'theory of existence'. In the context of sociology, it means that a researcher's choice of method is deeply connected to whether they think that the social world has an **objective reality** or whether it is constructed by human interaction. Sociologists therefore ask questions such as:

> **Key concepts**
> **ontology** the study of being and existence, and of the type and structure of objects
> **epistemology** the study of knowledge, including its scope, methods and validity
> **objective reality** a reality that is believed to exist independently of people and society

What is more important, society or the individual? Should research emphasise social structures or human **agency**? Should sociologists research macro (big) aspects of society or micro (small) interactions? There is no right answer to these questions: it is enough to acknowledge that society and self are intimately linked.

If a sociologist prioritises the idea that social structures shape human behaviour and our sense of self, they are more likely to adopt an **objectivist ontological position**. This means that they believe there is a real, external social world, containing 'things' such as social classes, genders, sexualities, the economy, or health, which can be studied objectively – from a detached and value free perspective.

If a sociologist prioritises human agency (the individual), they assume that our sense of being – our 'reality' – is subjective, and constructed with others. In this case, they adopt a **constructionist ontological position**. For example, these sociologists might be interested in how understandings of gender, health or social class are constructed and performed according to social context, culture and history.

Epistemology

A sociologist's ontological position is related to epistemology – the study of the origin, nature and scope of knowledge. For example, they ask questions such as: What counts as knowledge? What is it possible to know? What stands as data and evidence? Sociologists can usually be placed in different camps depending on how they think researchers should study the social world.

Below, we explore four different epistemological positions: positivism, interpretivism, **realism** and **constructionism**. These are not the only positions, but they will give you a framework for understanding and evaluating the sociological studies described in this book. Understanding the connections between the choice of method and the ideas and assumptions that drive these choices will enable you to be a better researcher, more able to appreciate and critique the methods you have at your disposal.

Note that we make a broad distinction between qualitative and quantitative methods in sociology. However, while each of these methods might be more suited to certain epistemological positions, there is not a definitive, fixed association between them.

A. Positivism

Sociologists who adopt an objectivist ontological position may be drawn to a positivist epistemology. Positivist sociologists aim to replicate the natural sciences by studying the social world as objectively as possible, free from bias or personal interpretation. Émile Durkheim is most commonly associated with positivism, but you do not need to be a follower of the theory of **functionalism**, which is associated with his ideas, to adopt this methodology. Sociologists in this camp tend to begin their research with a **hypothesis**, which they then **operationalise** (see Chapters 2 and 4 for more about operationalisation). For example, a hypothesis might be that parental social class predicts a child's educational performance.

> **Key concepts**
> **objectivist ontological position** a belief that the social world is something that can be viewed from the outside, i.e. that it is external to us
> **constructionist ontological position** a belief that the social world is constructed with others and can only be known partially and subjectively
> **realism** a philosophical perspective that focuses on revealing the hidden truths and causes behind social phenomena
> **constructionism** a social scientific perspective that focuses on how the social world/social reality is a product of shared meanings rather than being an objective entity
> **hypothesis** in research, a statement about the predicted outcome of an experiment
> **operationalise** to turn an abstract concept into something measurable

> **Key concepts**
>
> **correlation** the observation of a relationship between two variables, e.g. sociologists show that social class (variable X) is related to educational outcome (variable Y)
>
> **causation** the scientific belief that a variable ('x') directly impacts, and changes, another variable ('y')
>
> **sample** a small section of a wider population selected to research

To operationalise (measure) class, the researcher might look at the parents' occupations.

Using statistical techniques, positivists look for **correlations** between phenomena to confirm or falsify their original assumptions (hypotheses). If the hypothesis is upheld, and a correlation is found, positivists often infer **causation** – that is, they identify social causes to explain the phenomenon they are studying. As the social world is complex, most sociologists prefer to suggest correlation rather than causation. Testing the hypothesis that parental social class predicts a child's educational performance, researchers would be able to show a correlation between these two variables, but they would not go as far as to say that parental social class was the only causal factor in educational performance. Instead, it is widely acknowledged that parental social class has a strong *influence* on educational outcomes.

Positivists usually draw on quantitative methods. They try to secure large **samples** of respondents by using methods such as surveys (structured questionnaires) or by analysing official governmental statistics. The researcher in the positivist tradition believes that it is possible to always remain detached, and strives to do so.

In summary, sociologists who embrace a positivist epistemology assume that one can only make legitimate knowledge claims by using a scientific method to discern and measure 'facts' about the world. Facts about the world include things like social class and occupation, but also norms, values, customs and institutions such as religion, family, laws and marriage. Durkheim calls these **social facts**, and argues they are the proper subject matter of sociology. Social facts exist before we are born, persist after our death and constrain all that we do, and therefore they are researchable and measurable.

Émile Durkheim (1858–1917)

Durkheim was a French sociologist who studied the influence of social structures on human agency. He was particularly interested in using data to reveal social facts. His most famous illustration was suicide, which he argued was far from an individual and personal act and should instead be explained by the extent to which an individual is integrated into society or subject to control (regulated by society). He used official statistics to show that suicide rates varied by time and place, and were linked to factors such as religion.

> **Key concepts**
>
> **Verstehen** an empathetic understanding; being able to put yourself in someone else's shoes while researching them
>
> **unstructured interviews** interviews that take a relaxed, informal approach so they are more free-flowing and conversational

B. Interpretivism

In contrast, and inspired particularly by the work of Max Weber, interpretive sociologists take a constructionist ontological position. They see the world as constructed by and through human interaction, and argue that the proper subject matter of sociology is the meanings that humans construct and share about the world. For example, they would be interested in the social rituals that people enact and the beliefs that they hold.

Interpretivists believe that the best way to secure this knowledge is through a deep empathetic and contextual understanding – what Weber termed **Verstehen**. There is no expectation that the researcher should be completely detached; interpretivist sociologists believe that they need compassion and understanding to study the world, and an ability to see it through the eyes of their subjects. However, they are still committed to objectivity, and must be ready to suspend their personal and subjective views to objectively study the world of others.

Researchers who take an interpretivist approach tend to favour qualitative methods, such as **unstructured** and **semi-structured interviews** and **observations**, as these are more likely to elicit the voice and experience of respondents. Sample sizes tend to be much smaller than in positivist approaches, and researchers focus on a detailed, in-depth study, with an interest in words and meanings rather than numbers. Examples of sociologists that use this type of approach are found within **symbolic interactionism** and **ethnomethodology**.

Interpretivists may have some ideas before they begin their research, but they are not interested in testing hypotheses. Instead, they collect data and use these as the basis for developing theoretical concepts and ideas. This is known as **grounded theory** and it takes an **inductive** approach – that is, it develops a theory from **empirical data** (data gathered from research), from the ground up.

Max Weber (1864–1920)

Weber was a sociologist of Prussian descent and is often described as a 'micro sociologist'. He was interested in how individuals interact with one another and create shared social meanings about the world. Weber recognised that these shared meanings in turn shape individuals' actions and choices.

C. Realism

Like positivists, realists assume that there is an external reality that can be studied, but they suggest that this is often hidden and that different methods are required to reveal it. Realists theorise that social structures such as capitalism, **racism** and patriarchy shape people's lives in real ways, but their discriminatory impact is often concealed. Theories such as feminism, Marxism and **critical race theory** try to reveal these hidden structures through their social research methods. These three key conceptual perspectives are explored below.

Realist researchers are influenced by both objectivist and constructionist perspectives. As such, they use qualitative and quantitative methods or mixed methods (see Chapter 9). Many realist scholars believe that while social structures shape people's lives, these structures are in fact products of human-made history and only exist because of that.

Key concepts

semi-structured interviews interviews that make use of a guide that outlines an order of open and closed questions to ask participants; within the questions are probes that ensure clarity of response and allow the interview to develop in different ways

observations the act of watching other people as they go about their everyday lives; observations can be covert (where the researcher is unknown to participants), overt (where the researcher is known), participant (where the researcher is present in the setting) or non-participant (where the researcher watches from a distance)

symbolic interactionism a micro theoretical perspective that focuses on how individuals make sense of their world, and how an individual's sense of self is produced by interacting with others and understanding how they are perceived by others

ethnomethodology the study of the methods people use to make social order; how order is produced and maintained

grounded theory a form of inductive research that generates theory out of empirical data

Key concepts

inductive describing theories that are built from the ground up – the belief that scientific laws can be inferred by looking at observable (empirical) social facts, from which theories can be built about how the social world works

empirical data data that is gathered through research methods such as observation, interviews or surveys

Karl Marx (1818–83)

Marx developed what has become known as a grand theory of society. Put simply, he argued that changes to the economy could and should explain all other changes within the social world. For example, he showed how class relations changed according to how the economy was organised. Under **feudalism**, the major social division was between lords and serfs, and under capitalism new social classes emerged (the bourgeoisie and the **proletariat**). Marx used historical material to chart the impact of economic change on the social world. He was especially interested in revealing social inequalities and their injustice.

I Feminist perspectives

Many (although not all) feminist scholars can be associated with the realist epistemology, as they focus on revealing how patriarchy structures the life chances and lived experiences of women (and men). A feminist realist epistemology studies the influence of patriarchy on people's everyday lives and its influence on sociological research. Feminists suggest that science is a masculine enterprise and therefore asks questions that are shaped by male interests. Indeed, in the 1960s sociology was described as 'gender-blind' because it ignored women and women's interests. A feminist epistemology thus aims to develop female 'ways of knowing'.

Ann Oakley (1944–)

Famously, in 1974 Ann Oakley noted that female-dominated tasks and practices, such as housework, had not been subjected to sociological enquiry. She also suggested that women were often invisible in discussions of social class, as they were usually assigned their husband's occupation in research studies. A feminist epistemology, therefore, does not just demand that different questions are asked and that patriarchy is revealed, but is also associated with different research methods. Again, Oakley was pioneering in this regard. She argued that many research methods, including interviews and surveys, were exploitative, because researchers used respondents to generate data without considering how the respondents might be affected by the process. As researchers strive to secure objectivity, they remove themselves from the research process, which can leave their research subjects alienated and unsupported. In her study of motherhood, *From Here to Maternity* (1979), Oakley deliberately chose to engage with her research subjects about pregnancy and child-rearing, aiming to always give advice and answer their questions rather than just extract information from them (see Chapter 5).

Overall, feminist methods tend to be qualitative and are designed to be participatory, in that the researcher and the researched are in dialogue and support one another. By recognising that researchers are studying the lives of people who may already be marginalised in society, feminist scholars often take a **standpoint** position. They recognise that their research is derived from a set of political values and seeks to bring about change.

II Marxist perspectives

Building on the work of Karl Marx, a Marxist realist epistemology focuses on how the social world reflects the way that the economy is organised. In other words, Marxists argue that societies are fundamentally shaped by capitalist relations of production. This is referred to as **materialism**, or a materialist theory of knowledge. Like the feminists who seek to reveal the impact of patriarchy, Marxists aim to uncover the impact of the often-hidden influence of capitalism, and they too are interested in research that is radical and agitates for change. To this end, they draw on both qualitative and quantitative methods. In the latter, they analyse official statistics, but rather than take the data at face value, they draw attention to the ways that statistics support the status quo and thus may not be an accurate documentation of the world. For instance, Marxist scholars might look at crime statistics to draw attention to the absence of reporting about and investigation of **white-collar crimes** (the crimes of the **bourgeoisie**).

III Critical race theory

Sociology has often neglected to engage with a full and proper examination of the sometimes-hidden impacts of racism and colonialism not just on life chances and inequality, but also on the theory and methods that sociologists use. In other words, sociology has historically been 'ethnicity-blind'. The long-term omission of W.E.B. Du Bois as one of the founding fathers of the discipline, and the continued marginalisation of authors and researchers of colour, are examples of the ways that sociology itself can be critiqued (see Chapter 4 for a discussion of Du Bois' research).

Critical race theory (CRT) seeks to challenge the prevailing thinking and research in sociology to show that important questions and ideas have been systemically **silenced**. Once again, the theory calls for sociological research to abandon its claims to be neutral and objective, and instead to be proudly and transparently shaped by politics – in this case the politics of '**race**'. This does not mean that objectivity is abandoned, but that the values that underpin research questions are acknowledged.

To decolonise sociology, scholars have created new methodological tools, emphasising storytelling and **discourse analysis** (see Chapter 7). The intention is to give voice to indigenous knowledge, language and cultures that, in the past and even today, have been silenced or misrepresented. It endeavours to show how racism operates and to bring into full view those questions that the White majority of sociologists have not noticed or been interested in pursuing. For example, this approach critiques ethnographies that study people from 'other' cultures, using questions and frames of analysis from a perspective rooted in the **Global North**. Ethnography has been referred to as a 'dirty' word (Smith, 2013) when those with power interrogate those without it. Observing and **othering** cultures voyeuristically, and studying those who are powerless, was often done without any reflexivity at all. For example, in many early anthropological studies, researchers used a western lens to judge the people they studied. Malinowski (1929/2012) wrote a book entitled *The Sexual Life of Savages in North-Western Melanesia* – a title that is derogatory. Indeed, the lenses that the anthropologists used dehumanised and exoticised the people being studied, constantly judging their actions and practices in relation to a perceived standard of 'western' normality.

W.E.B. Du Bois (1868–1963)

Du Bois was a contemporary of Marx, Durkheim and Weber, but is often overlooked as a founding father of sociology. He is credited with drawing attention to the pervasive reality of racial inequality, segregation and discrimination that characterised US society during his lifetime. Du Bois argued that the 'most significant problem of the 20th century is the colour line' and that this was the proper subject matter for sociology. Du Bois used his data to reveal the hidden injustices of 'race'.

Thinking further: Intersectionality

Some scholars have argued that it is not enough to focus on one standpoint, but instead researchers should endeavor to reveal how multiple forms of social injustice operate. Kimberlé Crenshaw (1959–), an American civil rights activist and critical race scholar, coined the term **intersectionality** to capture the combined influence of social inequalities – specifically, ethnicity and gender. Patricia Hill Collins (1948–), a Black standpoint feminist, also challenged feminisms' focus on White, middle-class women and urged feminists to look at intersecting oppressions which overlap and create a **matrix of domination**. She urged sociologists to incorporate the perspectives of women of different social locations including ethnicity, social class and sexuality. The work of global southern feminists such as Chandra Talpade Mohanty (1955–) has also been incredibly important in developing an understanding of the influence of colonialism on life chances and lived experiences.

Key concepts

discourse analysis the contextual examination and deconstruction of meanings conveyed in text and verbal exchanges

intersectional describing an analysis that tries to measure the combined influence of socio-demographic variables such as ethnicity, class and gender on life chances and experiences

Michel Foucault (1926–84)

French theorist Michel Foucault was particularly interested in social order and control. He argued that power was not held only by a government or state, but that it was diffused, from the bottom up, through all social relationships. Thus, rather than locating power in institutions led by powerful elites, Foucault said that power operated through dominant forms of knowledge produced in society, which he called **discourses**.

> **Key concepts**
>
> **documentary methods** methods that use pre-existing archives, such as newspapers, films, diaries, social media and government policies, as data that can be analysed

D. Constructionism

Constructionism is both an ontological and an epistemological position. Like interpretivists, constructionists believe that the social world is produced through human meanings and actions. However, constructionists place more emphasis on how social structures, cultures and ideas shape the very possibility of knowing about and experiencing the world. Constructionists are more interested in power and in revealing which social constructions of reality are dominant.

Postmodernism and **post-structuralism** are two sociological perspectives that adopt these epistemological assumptions. Michel Foucault was influential in developing these theoretical ideas (see Chapter 7). He believed that there were dominant forms of knowledge in society (**dominant discourses**) that shape the ways that people think and act. For example, medical knowledge is a dominant form of knowing about human bodies and disease. There are two main methods of documentary analysis: content analysis and discourse analysis (see Chapter 7). Constructionists use discourse analysis. They ask, for example: How did medical knowledge claims become dominant? How did other ways of knowing about the body and disease (such as homeopathy) become sidelined? Whose interests are served by these forms of knowledge?

Documentary methods drawing on discourse analysis are a central research method of sociologists working within this epistemological position. Specific forms of discourse analysis include media analysis (including social media analysis) and policy analysis.

The advantage of different epistemologies

The four epistemological positions discussed here (positivism, interpretivism, realism and constructionism) shape the way that sociologists study the world. Of course, each epistemology only gives a partial insight, and may blinker the sociologist in that tradition to other ways of seeing and studying the world. Therefore, although these four approaches do not always agree about the types of questions that should be asked, or the methods that should be used, it is useful to have multiple epistemological positions. Holding up different lenses to the social world enables sociologists to capture the complexity of society and gain a greater understanding of the world.

> **Question**
>
> Think about the broad topic area of educational inequality. What different types of research questions do you think positivists, interpretivists, realists and constructionists might want to ask? Which methods would they prefer? What sorts of data would they use? What insights might they generate? Do you have a preference for one approach over another?

Can sociology be considered a science?

Having different perspectives means that there are disagreements about whether sociology can be considered a science. In particular, sociologists debate whether sociology can really claim the scientific gold standard of objectivity.

Positivists like Durkheim think that social research can replicate the natural sciences. By contrast, interpretivists do not believe that social research can be scientific or produce objective facts about the world. While interpretivists aim to be value-free and remove bias from their research, they recognise that their own positionality shapes research questions and analysis.

Realists sit somewhere in between. They see sociology as a science – as the study of an objective reality – but they focus on revealing hidden structures, not just on observable phenomena. Because realists are often politically driven, they do not claim to be detached; in fact, they are often open about their bias and motivations.

Constructionism is even more reflexive about sociology's claims to be scientific. They regard all knowledge as socially constructed – produced in social and historical contexts. Therefore, constructionists do not view sociology as a science and are careful about making claims to 'truth' and 'fact'.

Deciding whether sociology can be considered a science demands more than a simple weighing up of these four positions, and several key philosophers of science offer deeper insights.

Popper's 'falsification'

Karl Popper (1935/2002) argued that the claim to science was often judged on the wrong criteria. Rather than focusing on trying to prove a theory, science should instead try to *disprove* it, eliminating the possibility of all contradictory evidence. Popper called this **falsification**. He therefore disagreed with Durkheim, who believed that the standard for science was met if a hypothesis could be proved. This is what Popper called the 'problem of induction' – while inductive research helps to build theories of knowledge by drawing general laws from empirical observations, we cannot be certain that this knowledge is true.

Popper's theory of falsification suggests that scientists must look for contrary rather than supportive or verifiable evidence. Thus, we might ask whether sociology could be considered a science. Popper was critical of Marxism, arguing that it lacked scientific status as the theory kept being adapted to make sense of contrary evidence rather than simply being rejected. Popper believed that science can only progress through the process of falsification and the elimination of unsupportable ideas. He did not say that unscientific theories (like the theoretical perspectives within sociology) were not enlightening, simply that they did not meet criteria to qualify as proper science.

> **Key concepts**
> **falsification** the idea that in order for a theory to be considered scientific, it must be possible to test it and (conceivably) prove it false, e.g. the hypothesis 'all swans are white' can be falsified by observing a black swan

Kuhn's paradigms

Thomas Kuhn (1970) argued that all members of scientific communities work within a framework (**paradigm**) that shapes the type of questions they ask and blinkers them to other ways of seeing the world. Scientists working within a particular paradigm engage in periods of 'normal science'. As science progresses, paradigms shift, but there is always one dominant paradigm at any point in time. For example, a paradigm shift occurred when Einstein's theory of curved time and space challenged Newton's force of gravity theory. Kuhn would argue that sociology is not a science because it is not framed by one unifying paradigm.

Nevertheless, two important points about the status of sociology can be drawn from Kuhn's work. First, the very fact that sociology has many paradigms means that sociologists are constantly holding knowledge to account, rather than being bound and blinkered by one dominant set of ideas. Second, his idea of paradigm shifts helps sociologists to consider the discipline reflexively and to recognise that some sociological ideas might need to change. For example, feminists, Marxists and critical race theorists have all challenged the theory of functionalism, and functionalist ideas are becoming increasingly redundant in sociology.

Lakatos's research programmes

Imre Lakatos (1978/2001) merged the work of both Kuhn and Popper to create a hybrid account of what stands as science. While he agreed with Popper's idea of falsification, he did not think that contrary evidence meant that a theory had to be absolutely rejected. Instead, Lakatos described **research programmes**. These groups of scientists worked with 'hard core ideas' that were deemed irrefutable, but which had 'auxiliary hypotheses' (additional hypotheses) that could be tested and falsified.

Overall, Lakatos's work helps us decide whether disciplines employ good or bad science. A good science research programme delivers new predictions and new testable ideas. It can and must be continually tested.

> **Key concepts**
> **research programme** a set of theoretical ideas that are core to the workings of a group of scientists; if these theoretical ideas are abandoned, then the whole programme must be abandoned

> **Thinking further**
>
> Lakatos's work can be linked to sociology using the example of feminism. Feminists would argue that patriarchy is irrefutable (the hard core idea). They are reluctant to give up on the idea of patriarchy, but it is impossible to entirely verify or falsify this powerful structure. However, it *is* possible to test various aspects of patriarchy through a set of auxiliary hypotheses. For example, feminists could test the hypothesis that women do more housework than men.

When feminists locate some contrary evidence, or if a prediction turns out to be false, it does not mean that the whole research programme of feminism and its critique of patriarchy is redundant. For instance, women's earning power may now be equivalent to men's in some jobs, but this does not mean that patriarchy has been overthrown or is no longer a power structure. Such falsification does not mean that feminism is unscientific.

Feyerabend's contribution

Paul Feyerabend (1975/2010) believed that science is conducted according to a dominant paradigm. However, unlike Kuhn, he argued that science has been straightjacketed by the dominant paradigm. He suggested that the best way of being 'scientific' was to draw on multiple methods rather than restricting science to a particular methodology.

Where does this leave sociology?

If we follow Kuhn, sociology is *not* scientific because it has multiple paradigms operating at once (rather than one replacing another as a course of 'progress'). However, for Feyerabend, having multiple paradigms is something that characterises good science.

If we follow Popper, sociology fails the test of science because it does not fully engage in the process of falsification of its ideas. In contrast, Lakatos would regard some sociology to be scientific because the hard core ideas can still be used to predict the social world, even when there is contrary evidence.

Overall, it is safe to say that sociological research cannot ever be value-free, but this does not relegate its work to the domain of non-science. Rather, sociology can make claims for objectivity and rigour. In addition, by reflexively recognising its own biases and limitations, sociological knowledge has carefully considered 'social scientific' value.

3. STRIVE: Representativeness

Sociological research is not just common sense, because it never simply relies on the lived experiences of one person or one small group of people. Rather, sociological research strives to ensure that the evidence and data gathered represent the entire **population** being studied. Think back to your conclusion about whether or not Stacey Dooley's research (her investigative journalism) into social issues was sociological. Speaking to a handful of people about their experiences may prove insightful, but it cannot claim to represent the voices and experiences of everyone. Although social researchers face the same challenge, proper social research uses rigorous techniques to try to achieve 'representativeness'. This means that to be judged as sound, research must show that its conclusions are applicable to people and groups beyond those who have been directly studied, and those that experience different circumstances. To achieve this, sociologists use a process called **sampling**.

Key concepts

population the group of people that a sociologist wants to know about in their research

sampling a method involving the study of a subset of the population under investigation

To be confident that your research is representative, you would ideally conduct your study with everyone within the population that you are researching. For example, if you wanted to research undergraduate students in the UK, your population would be all students enrolled on an undergraduate course. Realistically, of course, it is often impractical – not to mention expensive – to include every person in a population. Nor is it necessary. Instead, researchers select a sample from their chosen population in the confidence that the results will reflect the experiences of the wider group. So, how do you select your sample? Who should be in the sample? And how big should the sample be? There are multiple ways of choosing a sample. Here, we distinguish between two broad methods – probability and non-probability.

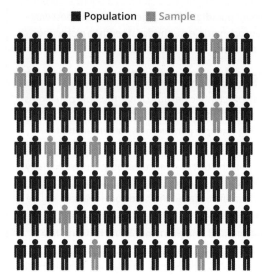

Key concepts

sampling frame the full list of people that a researcher wants to study, from which a sample is drawn

stratified sampling when populations are divided into groups for the purpose of sampling, usually by demographic characteristics such as age, ethnicity and social class

random/probability sampling a sampling method whereby everyone in the population has an equal chance of being selected

A. Probability sampling

Probability sampling is designed to ensure that the sample is as robust and representative as possible. The best way to achieve this is if the researcher knows all the people in the population that they want to study. In this way, they can produce a definitive list of the population to sample from.

For example, if you wanted to interview university vice chancellors, it would be easy to find a list of the names of all vice-chancellors and then create a **sampling frame** from which to select subjects for the research. You might decide to sample randomly from this list by pulling names from a hat. However, if the list is longer (i.e. the sample frame is bigger), you might use a computer to randomly select the sample. You could be even more systematic by taking every third or tenth name on the list, depending on how long it is and how many people you want to include. You might also decide to ensure your sample is representative of the gendered population of vice-chancellors by dividing them into groups by gender and selecting names from each group. This is called **stratified sampling**, and can include grouping the population by age, ethnicity and social class, for example.

Each of these methods is **random sampling** (also known as probability sampling), because there is a chance that anyone within the sampling frame could be chosen. The researcher is not deciding who is included in the research.

Random sampling is the most representative social research method and is favoured by positivist sociologists. It is the only way that researchers can be certain that their research is **generalisable** (applicable to the wider population). This is also sometimes referred to as having external validity.

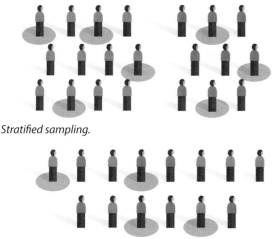

Stratified sampling.

Simple random sampling.

B. Non-probability sampling

Often, it is not possible to identify an entire sampling frame. For example, if you were interested in finding out about educational inequality and the use of private tutors in state education, there is no official list of students that have this advantage. Therefore, social researchers need to use other mechanisms to find a sample. One option would be to use a **convenience sample**, which involves drawing a sample from people that you know or have easy access to (for example, if you were a teacher you could draw on students in your own class).

A **snowball sample** is a type of convenience sample in which an initial contact leads to further possible respondents. For example, if you were investigating educational inequality and the use of private tutors, you could contact a private tutor and ask them to introduce you to some of their students. This type of sampling method often depends on a **gatekeeper** – a person or institution that gives you access to your respondents. The method is often used when studying vulnerable groups or sensitive subjects, particularly criminality.

This type of sampling method is called non-probability sampling, because there are no controls over who is being researched and not everyone in the population being studied has an equal and probable chance of being selected for study. It is also known as **non-representative** or **purposive sampling**.

> **Key concepts**
> **convenience sample** a sample drawn from an easily accessible population
> **snowball sample** a type of convenience sample, in which an initial contact leads to further respondents, like a snowball building up more snow as it rolls
> **gatekeeper** a person who can help a researcher gain access to a field
> **non-representative/ purposive sampling** sampling that is not random and so not everyone has an equal chance of being selected for study

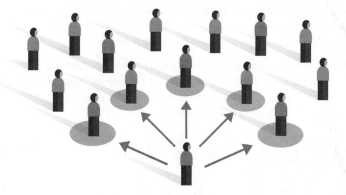

Convenience sampling.

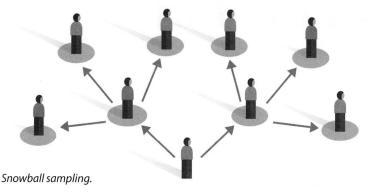

Snowball sampling.

Key concepts

quota sampling
an attempt to balance a sample when using non-representative methods by including particular demographics; quota sampling is not done as systematically as probability sampling

Question

What are the advantages and disadvantages of probability and non-probability sampling? Think about representativeness, cost, ease of access to the population being researched, time, etc.

There are other ways that social researchers try to secure a non-representative sample. **Quota sampling** is a strategy that researchers use to select more representative groups. For example, to balance the research, you might try to ensure that you cover a whole range of different ethnicities in the student sample being studied. Quota sampling is like stratified sampling in that it aims to cover different socio-demographic groups; however, quota sampling does not have a systematic method for choosing those people and therefore cannot be said to be truly representative.

Does this mean that non-probability sampling methods are equivalent to journalism? Absolutely not. While the researcher may not be able to claim full representativeness, their research process ensures that there is transparency and reflexivity in deciding who is selected to be studied. The fact that researchers detail their process means that anyone reading the work knows exactly how the sample was chosen and who was or was not included. In addition, areas of vested interest or researcher bias are openly acknowledged and the limitations of research are clearly outlined.

C. Sample size and response rate

Social researchers must decide how large their sample should be and whether they have contacted enough people. In quantitative research, larger samples are obviously more representative, but there is no formula for making the decision about how many people to study. Nevertheless, when undertaking surveys, for example, it is generally agreed that there is no need to go beyond 1000 questionnaires – more than this would not necessarily yield any further important results. Social researchers also need to consider aspects such as research cost (which is often constrained by the funding they have secured), access to the population being researched, and time available for research, which all constrain the number of people that can be included.

In contrast, qualitative researchers, who use interviews and observations, often use the term **saturation** to describe the point at which they have gathered enough data. Again, there is no definitive number of interviews that ought to be carried out, but it is generally agreed that saturation has been achieved when interview data begins to repeat the same information and no new meaning is being elicited.

Researchers must also be confident that enough people have taken part in their research. This is called the response rate. Sample size is compromised if people refuse to participate, as this can impact representativeness. Quantitative researchers generally consider a 50% response rate acceptable, but in practice the response rate is often much lower.

> **Key concepts**
> **saturation** a term used to describe the point in the research process when no new information is being found and the researcher can assume that data collection is complete

4. STRIVE: Impactful

Good social research should not simply research the lives of people; it should also aspire to enact meaningful social change. Plummer (2013) referred to this as a commitment to 'humane sociology', in which research questions are socially significant. For instance, while you might be interested in your aunt's allotment or your neighbour's chicken farm, you would have to ask yourself: *Is undertaking this research of wider relevance to society?* Sociologists must ensure that the methods they employ produce results and evidence that are useful to governments as well as to the people that they are studying. While it is not possible to come up with a checklist for measuring impact, it is important when reading research to ask the question: *Did this research make a difference?*

> **Activity**
> Taking the example of allotments, work in pairs to discuss and note down some ways that undertaking research into this area could be socially significant.

Because sociology studies the very lives that we live, inevitably its research has a transformative function – it unsettles ideas that we take for granted and impacts those who are studied as well as those who read the research.

5. STRIVE: Valid and reliable

The terms 'validity' and 'reliability' are critical for the evaluation of research.

Validity

A research study is valid when you are confident that you are measuring what you set out to measure. Therefore, a valid study gives a true picture of the topic under investigation. Asserting validity can be challenging for sociologists – social concepts are always complex and therefore researchers must rely on measures that cannot fully capture their essence.

For example, consider how you might measure something like social class. You cannot simply ask someone what social class they belong to, because you do not know what they understand 'social class' to be. Different people might use different sets of criteria to define their class – one person might think of their own occupation, while others might think of their parents' occupations. Some people might consider how much money they have, while others consider their educational achievements.

Social researchers try to ensure research validity by using **proxy measures**. For example, sociologists typically use occupation as a measure for social class. Because social class is not simply an economic calculation but is also about cultural values and social networks, some sociologists might argue that an occupational measure of class is not entirely valid. Chapter 4 reviews Bourdieu's (1979) conceptions of class and Savage's (2015) attempts to improve research validity in this area.

One way to enhance a study's validity is to use different methods to examine the same phenomenon, which can increase confidence that the results reflect the 'truth'. This process is described as **triangulation** and is a form of methodological pluralism – the combination of different methods to give a full picture of your topic. It is also sometimes referred to as using 'mixed methods' (see Chapter 9).

> ### Key concepts
> **proxy measure** a way of measuring social phenomena that cannot be directly observed by using a substitute or stand-in variable (e.g. social class is most commonly measured by occupation)
> **triangulation** using multiple methods, theories or researchers to improve the reliability and validity of research

Reliability

All research, whether it is hard science or a social science like sociology, endeavours to produce robust findings. For research to be reliable, its instruments must be consistent, measuring the same phenomena and producing similar results at different points in time and when used by different researchers. Sometimes, then, reliability can be understood as repeatability: if you or someone else repeated the study, could you be sure that you/they would get the same results? If you can be confident that the research is repeatable, then it is reliable.

> **Thinking further**
>
> The UK Census is generally regarded as highly reliable: it is made up of questions that have been carefully tested; the researchers are trained on how to ask the questions and record the responses; and it has a huge response rate (97% households). In fact, there are government fines for people who do not complete it, making it hugely representative. It has also been successfully repeated since 1801.

Case study: Judgment criteria in qualitative research

Some social researchers have argued that the positivist scientific criteria of validity and reliability are not suitable for judging the rigour of qualitative research informed by other theoretical frameworks, such as interpretivism or constructionism. Lincoln and Guba (1985) suggest alternative 'trustworthiness' criteria:

Quantitative	Qualitative
Internal validity (truthfulness)	Credibility
External validity (generalisability)	Transferability
Reliability	Dependability
Objectivity	Conformability

- Credibility refers to whether the researcher has spent significant time in the research setting and has used different data sources, methods and researchers (triangulated).
- Transferability refers to how much detail the researcher gives about their research study – what they found, where it took place, who they spoke to, the wider context, etc. – so that any reader can appreciate the insights and transfer them to other settings.
- Dependability refers to being transparent about what methods have been used to reach conclusions and the logic behind the process undertaken.
- Conformability refers to the process of reflexivity, transparency around biases, and steps taken to eliminate bias. These steps could include, for example, peer debriefing, in which a researcher asks another researcher to explore possible biases in the research data or findings, or member-checking, in which transcripts or field notes are given back to respondents to check for accuracy.

Lincoln (1995) later revised these criteria. She suggested that qualitative research should not have to parallel the existing quantitative metrics but should instead be evaluated according to whether the researchers could show that they had:

- taken into account the research values of the community being studied
- acknowledged their own author's position
- given true acknowledgement of the participant's voice
- been reflexive
- built reciprocal trusting relationships with their participants.

6. STRIVE: Ethical and reflexive

Social researchers must be mindful of practical ethical issues. Additionally, they should reflexively consider the influences on, and impact of, their research.

Ethical

On a practical level, sociologists abide by ethical guidelines established by their national sociological associations.

Research tip

British sociologists conform to the ethical guidelines from the British Sociological Association: https://www.britsoc.co.uk/

Key concepts

deception when a researcher withholds information from, or intentionally deceives, respondents about the true nature of a study or about the researcher's true identity

informed consent agreement to participate in research in full knowledge of the study; some vulnerable populations are unable to give informed consent, such as children or people with learning disabilities

Becoming a social researcher

The following questions are helpful in assessing the ethical standing of any research study:

- Is the research necessary?
- Is the researcher qualified to undertake the study?
- Is the research likely to cause any harm to any participants? For example, might the questions trigger memories or deal with sensitive subjects? Can these risks be mitigated?
- Has the research been properly explained to research participants, and do they understand what will happen with the results, ensuring no **deception**?
- Have all participants given their **informed consent** to be involved in the research? (This is sometimes more difficult to achieve if the researcher is undertaking covert observations or when dealing with vulnerable populations such as children or people with learning disabilities.)
- Do all participants know that they can withdraw from the study at any time?
- Will all participants be anonymised in any results? If not, do participants understand that their identity might be revealed?
- Will research be stored safely, with access to data protected?

Thinking further: Decolonising ethics

Research ethics typically focus on identifying sets of rules that researchers must follow to protect themselves and the people they are researching. While useful, these ethical rules are arguably a product of neoliberal, Global North concerns that focus on individuals, and aim for objectivity and detachment. However, we can draw lessons about how to engage in ethical research practice from global southern philosophies and perspectives such as Ubuntu, Sikhism and Indigenous knowledge.

For instance, scholars can draw on the African conception of Ubuntu to guide their research. Ubuntu is often translated as 'I am because we are' and is focused on shared humanity and community. Ubuntu research ethics would therefore aspire to emphasise common good, human dignity, mutual respect, tolerance, and reciprocity. Research can be judged on the extent to which it contributes to positive, humane relations, the extent to which it reveals and challenges oppression and injustice, and the extent to which it promotes peace and consensus while embracing and accepting diversity (Ujomudike, 2016).

Research ethics can also be shaped by religions such as Sikhism. Sikh ethics emphasise social justice, tolerance and service to others. From this perspective, research ethics should be framed around kindness, understanding and reflexivity. Here responsibility cannot be reduced simply to questions of objectivity and consent but demands an appreciation of the impact of research and a commitment to service to others and humane and impactful enquiry (Singh, 2023).

There is also a recognition that many indigenous peoples have been othered and harmed by research from the Global North and therefore research should be judged according to whether it is safe, empathetic, kind and respectful and whether it builds tolerance and good relationships. In this regard, an Indigenous ethics would similarly place value on trust, honesty and humility, with a commitment to research that promotes social justice, care for all, as well as a sustainable respect for nature and the land (Lovo et al, 2021).

Reflexive

Even when research is deemed to be practically ethical, it is not necessarily value-free. Researchers may strive for objectivity, but they cannot always eliminate bias. They must therefore be transparent about their own position, recognising the influences that shape their work. It is always helpful to ask questions about the positionality of the researcher(s) when reading their work. As you progress through this book, you will want to keep this question to the forefront of your mind, in order to become analytical and critical in your reading of research studies. Remember that positionality is shaped by gender, class, ethnicity, sexuality, disability and age, as well as ontological and epistemological preferences. Moreover, if sociological research is funded by governments, charities or private companies, the interests of these stakeholders should be acknowledged.

You can even ask questions about the positionality of the discipline of sociology. For example, acknowledging the absence of work on the impact of colonialism reveals that much sociology has been conducted from the perspective of the Global North. As such, many research findings assume universality without exploring the importance of studying the social world from all perspectives.

Meghji (2021) argued that many studies assume that their findings apply universally. Meghji critiqued Garfinkel's breaching experiments – which explore what happens when social norms are broken – for failing to appreciate how people from the global majority might experience life very differently from the White middle-class people who were Garfinkel's main respondents. Decolonising methods does not necessarily require a whole new set of research tools (although it might put more emphasis on storytelling, for instance). However, it does require researchers to be reflexive about the questions that they ask. It demands that they think about where, when and how data is collected, who it is funded by, and that they are careful about the ways in which they analyse and present it. Meghji cited the insights of Walter and Andersen (2013) on how to collect and interpret statistical data:

> ... while Western methodologies may use statistics to investigate questions pertaining to 'Indigenous enrolment at Australian universities', an indigenous statistics approach would instead look at statistics to look at 'how well do Australian universities incorporate indigenous participation'. (p. 141)

Being a good researcher means being aware of your positionality and its influence on the research process. It involves the ability to be reflexive. Social researchers must acknowledge the contextual factors that shape research agendas and research methods. They must understand, analyse and evaluate their personal values and beliefs, and the values and interests of any funders. They must therefore be mindful of the limitations of their research studies. Reflexivity requires an openness and an acceptance that the researcher is part of the research process rather than detached from it.

Because sociology studies all aspects of human life, research does not sit in a vacuum – or laboratory – separate from its research subjects. Rather, sociological research inevitably changes the very world that it studies. Researchers must be aware of three key consequences of research: the Hawthorne Effect, 'going native' (over-immersion) and the double hermeneutic.

The Hawthorne Effect

When a researcher undertakes observations or asks questions, they can have an impact on those being watched or questioned. This is referred to as the Hawthorne Effect (after sociology experiments carried out in the 1920s and 1930s in Western Electric's Hawthorne Works, USA), and is the recognition that individuals might modify their behaviour or responses because they are aware they are being studied.

'Going native' (over-immersion)

By undertaking sociological research, a researcher may become so involved in their work that they can lose sight of the research questions or so immersed that they fail to engage with their study with the rigour and objectivity that research demands. This is sometimes referred to as 'going native', although this term might be criticised for colonial bias. Nevertheless, the warning that researchers can become too immersed in their topic is an important one and the term 'over-immersion' can be used instead.

Double hermeneutic

Research findings are often read by non-sociologists such as journalists, politicians and ordinary people, which can have both intended and unintended consequences. Although much sociology *aims* to be transformative, it is also important to consider the *unintended* impact of research – sometimes referred to as the double hermeneutic.

Hermeneutics is a method that studies the ways that texts are interpreted. The term 'double hermeneutic' attempts to capture the process of what might happen when people read sociological research. Giddens (1992/2013b) described how sociological concepts, ideas and findings often leak into everyday speech and understanding. They circulate in the social world that sociologists analyse, and can change the experiences of those who encounter them. It is therefore imperative that sociological research is aware of all these unintended effects. For example, Giddens noted that when sociological research into divorce became widely read, the idea of divorce became more acceptable and less stigmatised, which arguably contributed to rising divorce rates.

You are now equipped with a set of rules to help you judge sociological studies. In the next chapter, we examine the steps you need to follow to undertake your own research.

CHAPTER 2: DOING YOUR OWN SOCIAL RESEARCH

One of the most exciting things about studying sociology is doing your own research. What do you want to know? How will you find out about it? To conduct your research rigorously, you must follow the STRIVE rules outlined in Chapter 1, but will you also need to carefully follow an established research process to formulate the questions you will ask, to choose the method you will use, and to decide which mode of analysis you will employ.

Doing background research and devising a research question

Well-formulated research questions take work to produce. They come from a systematic process grounded in careful background reading and a thorough consideration of what sociologists already know. This does not mean that your area of interest cannot be sparked by your own experiences, but you should not let your own biography define the questions that need to be asked. Usually, a research question emerges after a researcher identifies a broad topic of interest and analyses the literature within it. This is known as conducting a literature review.

Literature reviews are just what the name suggests: a review of the existing research literature in the field of interest. This might be journal articles, books or chapters of books written by academics, as well as policy documents and official publications from governments and other agencies. Usually, researchers will place some disciplinary boundaries around their literature reviews. For instance, if they want to know about gender and housework, they may begin by researching what sociology already has to say on this topic. They will search journal articles and academic books, and assess the existing research for the precise questions that have been asked, the theoretical orientation of the researchers, the methods that have been used, the sample that the researchers have focused on, and so on.

Analysing existing academic literature allows researchers to:

- understand what ontological and epistemological orientations have already shaped this area of enquiry. What sort of questions have been asked? Were any questions *not* asked?
- understand what methods have already been used to examine the topic. Has there been a bias towards one method over others? Might other methods offer different insights?
- collate the empirical evidence that already exists to see if there are any gaps, and to gauge whether the evidence is contemporary or dated. Does the research need repeating? Are any questions left unanswered? Have the researchers identified limitations in their own studies or suggested further research that is needed?

These three steps will help you to devise your own question, which in turn must be carefully conceived. A good research question is focused, clear and, importantly, researchable.

Of course, you may find that your area of interest has not yet been adequately explored within the field of sociology. If so, you may have to venture into other disciplinary areas to find out what research exists. In this case, it is important to be attentive to the critiques that sociology has levied at these disciplinary positions. The Researcher insights box gives two examples from our own research.

 Researcher insights

When Sarah (2018) wanted to understand why there were increasing rates of mental illness among students, she found that the existing sociological research had not examined this area critically, but instead that psychology dominated enquiry. When Jennifer (2016) wanted to research assisted dying, she also found that the field of sociology had very little to say, and that most research in this area was from the disciplines of law, philosophy and medicine. In cases like this, the options are to either broaden or narrow the literature searches, depending on the research already available in sociology.

 Thinking further

You can visualise the process of conducting a literature review as an inverted triangle. Consider the following example based on the broad topic area of gender and housework.

- What research already exists? Has the research looked at other intersections with gender in more detail, such as sexuality, ethnicity, social class, occupation, disability, age and geographical location?
- Has the research looked at different types of housework, historical changes in housework, changes in marriage and divorce patterns, whether families with children have different divisions of housework?
- Has the research been dominated by a particular methodology or theoretical orientation (what concepts are being discussed in the field)?

Once you have detailed the research that has been conducted, gaps will be revealed and your own questions and objectives can be clarified.

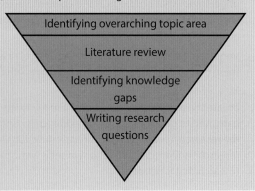

Clarifying any gaps in the existing research and your own research objectives is not the final step, however. Next, you need to carefully formulate your research questions before establishing the method you will follow to gather data. But what does a sociological research question look like?

> **Thinking further**
>
> Some researchers, including Denscombe (2010), argue that social research questions can comprise:
>
> - descriptive questions, in which researchers describe a phenomenon (e.g. *Who does most of the housework?*)
> - predictive questions, in which researchers explore whether something happens under different circumstances (e.g. *Did men do more housework during the COVID-19 lockdown?*)
> - explanatory questions, in which researchers see if manipulating variables has different consequences (e.g. *Does the share of domestic labour change when a couple has a baby?*)
> - evaluative questions, for example assessing whether a variable has the benefits it is said to have (e.g. *Did the Equal Pay Act impact the domestic division of housework?*)
> - transformative and empowering questions – that is, whether something can be improved or whether lives can be enhanced (e.g. *Would wages for housework improve women's experiences of domestic equality?*)

Devising a research question

You can use the first two elements from the acronym STRIVE to frame your research questions.

To begin with, it is essential to keep the premise of sociology in mind when developing your questions. Good *sociological* questions explore how individual experiences are shaped by wider social contexts, social structures and discourses. For example, however interesting it might be, it would not be properly sociological to ask a question such as *How do pets enhance self-esteem?* because this question focuses on the inner life of the individual rather than explaining pet ownership in a broader social context.

The research question is also shaped by the *theoretical* disposition the researcher takes towards studying the world. Positivism, interpretivism, realism and constructionism all lend themselves to different types of questions. Typically, a positivist will ask quantitative questions, and interpretivists and constructionists will ask qualitative questions. Realists will often use a combination of both. This is, however, a crude distinction and will not always be the case.

Look at these examples of how different theoretical orientations in research might shape the questions sociologists ask:

- How many people own pets? (descriptive, quantitative, not particularly sociological)
- What sort of people own pets? Is there a relationship between social class and pet ownership? Is there a distinct difference in characteristics between owners of poodles and terriers? (quantitative, positivist)
- What are people's experiences of living with pets? (qualitative, interpretivist)
- Is there a relationship between pet ownership and loneliness? (quantitative, positivist)
- How do discourses of dangerous dogs reinforce class divisions? (qualitative, realist or constructionist)
- How does the media portray fox hunting by hounds? (qualitative, realist or constructionist)

Generally speaking, positivist quantitative questions ask *What is…?* or *How many…?* For example, *What is the relationship between x and y?* or *How many people…?*

Interpretivists' qualitative questions generally ask *How do people experience x and y?*

Realist questions tend to focus on revealing hidden meanings and structures, and are therefore framed as *How is x portrayed (in media or policy)?* or *What is the impact of x on y?*

Constructionist questions tend to ask about power relations – for example, *How does the discourse of x operate?*

Becoming a social researcher

Social researchers sometimes frame their research question as a hypothesis. Typically, hypotheses are used by quantitative sociologists who are trying to test theories. A hypothesis is a statement of what you expect to find out, which your research then tests. For example, a sociologist might construct a hypothesis about the relationship between parental occupation and educational outcome as follows: *Children from middle class families do better in school examinations than children from working class families.*

Of course, qualitative researchers have inclinations about what they might find out through their research too. However, quantitative researchers will often write this down as a formal part of the research process, and will note in their findings about whether their hypothesis was right or wrong.

Activity

Think about the example of dog ownership. Look at these research questions.

- Is there a distinct difference in characteristics between owners of poodles and owners of terriers?
- Is there a relationship between pet ownership and loneliness?

Reframe each question as a hypothesis.

Operationalising concepts

As discussed in Chapter 1, *operationalising* refers to the process whereby a concept is made into something measurable (an indicator). It stands in for something else – a proxy measure. When considering the relationship between pet ownership and loneliness, for example, a social researcher would need to determine how to operationalise loneliness. Loneliness is defined by the British government as follows: *A subjective, unwelcome feeling of lack or loss of companionship. It happens when we have a mismatch between the quantity and quality of social relationships that we have, and those that we want.* (ONS, 2018). Because it is subjective, loneliness can mean different things to different people.

In the UK government Census, the questions asked to indicate loneliness include:

- How often do you feel lonely?
- How often do you feel isolated from others?
- How often do you feel you lack companionship?

The answer options provided are *hardly ever or never, some of the time* and *often*.

Activity

Here are some other questions about loneliness:
- How many times in the last week has someone visited you?
- How many times in the last week have you been invited to someone else's house?
- Do you have a partner, family or friend living with you?
- Do you have any children living with you?
- Do you have any pets?

What problems can you identify with these questions? For example, think about whether the questions could be interpreted in different ways, whether they are leading, whether they ask more than one question or whether the data generated from the answers would be easy to analyse.

Research tip

You can access the UK government Census at: https://census.gov.uk/

Choosing methods

Once you have decided on your theoretical framework and your research question, it is likely that the best method to use will be obvious to you. This is because your question and theoretical orientation will generally demand a specific method in order to capture data to answer your question(s) most appropriately.

Activity

Consider the following questions:
- How do parents socialise their children into different gender roles?
- How are single mothers represented in the popular media?
- Why are fewer minoritised ethnic groups enrolled in Russell Group universities?

- Why are there fewer women than men in senior positions in banking?

What methods do you think would best help you to answer each of these questions?

What different sets of data would each method give you?

You may have realised that to answer 'Why' questions, you will probably need to talk to people directly, so interviews and focus groups may be the best method.

If you want to explore whether there is a relationship, or even a predictive relationship, between two variables you will probably want to use quantitative methods, such as analysing official statistics. Alternatively, you may want to design your own survey.

If you want to know how something or someone is represented in popular media, then you will be more likely to choose a form of documentary method.

Once you have established your research method – whether you will undertake a survey, an interview, an observation, use documentary methods or official statistics, for example – you will need to devise your sample (see Chapter 1).

Research ethics

No research can take place before you have secured ethical approval. Even if you are undertaking a piece of research in school or university, it is essential that it is approved. **Ethics** often come in the form of checklists designed to ensure that the researcher has considered the core ethical questions and dilemmas. However, research ethics extend beyond this. Sociologists must consider questions of positionality and reflexivity, asking questions such as:

- What is my own positionality in relation to those I am studying?
- Is there a power relation between myself as a researcher and my research participants? How can I mitigate this power relation to the greatest possible extent?
- Do I have a particular bias towards those I am researching?
- If the research is externally funded, does the funder have a particular agenda or bias?

Questions

Should positionality stop a researcher from examining certain topic areas? Can reflexivity in research mitigate any positionality issues? For example, should racism only be researched by Black and minority ethnic researchers, or should White researchers conduct research into racism as a form of allyship? Should men interview women about domestic violence? Should women interview men about male infertility?

Becoming a social researcher

Sometimes, researchers recognise that their research is too problematic to conduct themselves, so they employ other people to do it.

Reflect on your own socio-demographic characteristics: your gender, ethnicity, sexuality, social class, age, (dis)ability, religious orientation. How might these characteristics impact your research? How might your own positionality enable your access to certain research fields? How might your positionality make access challenging or possibly unethical?

Carrying out the research

Gathering data and analysing results

Chapters 3–10 explore the different ways that social researchers gather their data. It is useful to distinguish between two main forms of data analysis to help guide you in your own choices.

Quantitative statistical analysis

Survey data and other forms of quantitative data (e.g. structured observations) can be subjected to statistical analysis. Broadly speaking, there are two kinds of statistical analysis: descriptive and inferential.

Descriptive statistics simply describe what the data says – for example, how many people own terriers or Dalmatians. Inferential statistics allow a researcher to make correlations, inferences or predictions – for example, they would enable a researcher to establish whether there is a relationship between ownership of certain dog breeds and social class. Inferential statistics usually rely on data manipulation and analysis through statistical software, such as SPSS and 'R'. You will often see the use of **standard deviation** measures in tables that show whether a relationship between variables is significant.

> **Thinking further**
>
> When conducting your own research, you may find it helpful to use a pilot study to test your methods before you complete a larger-scale version of the research. Usually, pilots are carried out for surveys that aim to capture a very large sample, but sometimes qualitative researchers will use pilots to test research instruments, such as interview schedules. Pilot studies enable the researcher to:
>
> - test their instruments – for example, to test the questions they are asking, to make sure they are understood by participants
> - test that their concepts are operationalised effectively
> - test any bias.

Key concepts

standard deviation
a statistical measure which refers to the extent to which data is spread out from the mean (average). It helps researchers decide whether the relationship between variables is due to chance or if there is a strong correlation. When looking at tables a significance level of 0.05 or less is the standard for suggesting that the results of correlation are not due to chance. It is derived using the following formula, which computer software calculates:

$$\sigma = \sqrt{\frac{\sum(x_i - \mu)^2}{N}}$$

Qualitative thematic analysis

Researchers using qualitative thematic analysis tend to generate and build theories through an inductive research process. Thematic analysis sometimes involves the use of software, such as NVivo, but typically researchers will read and re-read transcripts to look for commonalities within the data that form recurring themes.

Writing up your research

The final stage of the practical research process is drawing conclusions and writing up the research. How you choose to present your findings will also depend on your theoretical orientation and the norms governing your disciplinary field.

Positivists tend to follow scientific rules for writing up research. They usually report their studies in a procedural, systematic way that follows scientific research conventions.

 Thinking further

Typically, a scientific research study will follow this format:

- Abstract: a summary of the study, including key findings.
- Introduction: An overview of the study and an outline of the content of the written report.
- Literature review: A summary of the existing research, concluding with the research 'gap', research objectives and questions.
- Methodology: An overview of the theoretical approach being taken, the method used, the sample and sampling method, the judgment criteria (if qualitative), and the data analysis method.
- Findings: The **raw data** is either presented in quantitative formats, such as graphs or charts, or qualitative formats, such as snippets from interview transcripts. Quantitative studies usually have a separate findings section, which tends to be more descriptive than analytical. Qualitative studies often skip this section and present their findings in the same section as the discussion/data analysis.
- Discussion/analysis: More than mere description, in this section researchers draw out deeper analysis, usually presented as themes.
- Conclusion: A summary of the study, including limitations and directions for future research.

Key concepts
raw data data from existing datasets that has not been processed for use (manipulated or analysed in any way); sometimes called source data

Quantitative and qualitative researchers typically structure their research write-up in the same way, despite the significant differences in their approaches and theoretical orientations. However, qualitative researchers recognised that there was a problem with the traditional way that research was being written up, as it was usually presented in a factual, detached and objective way. Denzin (2002) and Denzin and Lincoln (2005) refer to this as a crisis of representation that occurred in qualitative research in the 1980s and 1990s. Using the traditional writing structure and detached style was problematic for many constructionist researchers (particularly postmodernists) because they recognise that all knowledge is socially constructed, and that research 'writing' creates a particular view of reality rather than objectively representing a true reality. As a consequence, researchers such as Richardson (1990) took a **narrative turn**, in order to bring the 'absent author' back into the writing process. This narrative turn challenged the traditional structure of presenting research in a way that was dry and detached and turned to more subjective and story-like ways of writing up research.

As you can see, even choosing a style for writing up your research involves decision-making and justification. This is something that you will have to reflect on carefully when you reach this point in your research.

In summary, this chapter has introduced you to the various steps involved in undertaking research. The next chapters outline a range of classic and contemporary research studies that have used different methods to investigate social phenomena across the areas of crime, education, inequality and family.

CHAPTER 3: OFFICIAL STATISTICS

What are official statistics?

Since the nineteenth century, there has been an increasing emphasis on collecting data about populations, generating a wealth of information about how people live their lives. Sociologists have reflected on this, referring to both the rise of a **surveillance society** and the creation of **data selves**.

Official statistics are any data that has been collected by the government, or by organisations acting on its behalf. These statistics are usually available to the public. Official statistics are one example of secondary data, and can be quantitative or qualitative. This chapter focuses on quantitative, official statistical data.

> ### 💭 Thinking further
>
> Statistics and qualitative data can be collected by bodies other than governments. Charities, businesses, schools, churches and many other organisations may hold data that researchers can access and use. Social media companies such as Instagram, Twitter and Facebook also compile datasets (big data), which can be useful for research (see Chapter 7). One example of qualitative secondary data is the Mass Observation Project, held by the University of Sussex. This is a national archive that collates open-ended answers to selected questions from hundreds of volunteers (referred to as observers), three times a year.

> **Research tip**
>
> You can access information on the Mass Observation Project at http://www.massobs.org.uk/

> **Research tip**
>
> You can access and download ONS data at the government website: https://www.ons.gov.uk.
>
> Some interesting and useful datasets available from the ONS include: the British Cohort Study; British Social Attitudes; census records; the Crime Survey for England and Wales; Family Expenditure Survey; General Household Survey; Labour Force Survey; National Child Development Study; National Food Survey; Opinions and Lifestyle Survey.

The Office for National Statistics (ONS)

In the UK, the government department responsible for official statistics is the Office for National Statistics (ONS). The ONS has data on a vast array of topics, ranging from the number of pets people keep, the hours people spend cleaning their homes, their beliefs about immigration, and how long they spend on social media, to large-scale studies on population size, health, crime, education, employment, income and religiosity. Sociologists use pre-existing statistics like these to examine many sociological questions.

Becoming a social researcher

If you are interested in data about a local area in the UK, such as a town or electoral ward, you can find statistics using the Nomis site (https://www.nomisweb.co.uk/). By entering your postcode, you can very quickly access information about the local population, including employment rates, how many people own or rent their homes, the social class and age of the population, as well as the types of qualifications that people hold.

Choose two different wards within the town in which you live and compare the data on employment and qualifications. You will likely find that within any town some wards are characterised by poverty and unemployment, while others are more affluent and highly educated. Young (1999) showed that such social polarisation often occurs in neighbouring wards situated, he said, 'cheek by jowl' and separated by a 'cordon sanitaire' – a major road, train line, river or canal, for instance.

 Statistics about housework

What can official statistics tell us about housework? The UK Household Longitudinal Study is a survey of 40 000 family households. In 2018, Kan and Laurie used this data to explore differences in who did the housework depending on the ethnicity of the families. The sociologists were interested in examining this variable because all previous research about housework had shown that women undertake the largest share of the work, but had not explored whether ethnicity had a separate influence.

Kan and Laurie took an intersectional approach, and the existing data allowed them to examine ethnicity as a factor even though it had not been the objective of the earlier studies. The database relied on self-reported ethnicities of heterosexual couples, which was then compared to the number of hours people spent on domestic labour, and their attitudes to work within the home. The researchers recognised that self-reporting of ethnicity may not be reliable, and that different communities may interpret what counts as domestic labour differently.

The research tested several hypotheses. The research team expected educational attainment, employment status and gender attitudes to vary by ethnic group, and that these variables would be correlated with the share of the housework undertaken by couples. Their analysis corroborated previous research, finding that women from all ethnicities did the bulk of the housework. However, the research showed that British Caribbean and British Indian men spent more time on housework than British White men. British Pakistani men reported the lowest share of household work.

From this brief description of the research and results, you can start applying the STRIVE criteria:

Questions

Look at the statistics about housework box. How important do you think it was to undertake this research?

How would you describe this research – is it an example of positivist, interpretivist, realist or constructionist sociology?

What do you think were the advantages of this method for this study?

What limitations can you identify in this study?

Sociological:
- By focusing on ethnicity, researchers made connections between broader social structures and individual experiences.

Theoretical:
- Like much research that draws on official statistics, this study can be described as positivist.

Representative:
- The dataset was large, so the researchers could be confident of its representativeness.
- However, only heterosexual couples were studied, making it less representative of the broader population.

Impactful:
- Much research is ethnicity blind, and this study provided important insights into the **domestic division of labour**.

Valid and reliable:
- The original research had been carefully devised by the government, increasing its reliability.
- Researchers had no control over the questions that were asked and, therefore, no control over how concepts such as housework were defined.
- The data relied on self-reporting of ethnicity. There may have been inconsistencies in how people defined their ethnicity, raising questions of validity.

Ethical and reflexive:
- The researchers did not have to gather the data themselves, which made the study cost- and time-effective, and the researchers' positionality did not have any impact. There were no ethical barriers to overcome.

Key concepts

longitudinal research research that involved repeating data collection over a period of time

Sociologists find statistical data both useful and interesting because it helps to build a picture of social life and to identify patterns, such as whether some groups are more advantaged and successful than others. The fact that governments have been collecting data for a long time also allows sociologists to examine the ways that society has changed over time. This is referred to as **longitudinal research**.

How do sociologists use official statistics?

Analysing statistics allows sociologists to address two fundamental sociological questions:

- whether societies are characterised by inequality
- whether inequality is increasing or decreasing.

Sociologists use existing databases to:

- **describe a given phenomena**, such as how many young people achieve GCSE qualifications at the age of 16 in any one year. Sociologists describe data in terms of numbers, percentages (the number of people per 100 in a population), and rates (often expressed as the number of people per 1,000 of the population).
- **show trends over time**, such as whether the numbers of people getting top GCSE grades is increasing every year
- **check if there are any correlations in the data**, such as whether high or low GCSE grades are more prevalent among certain groups. These correlations may be identified from existing datasets, but researchers can use the raw data and manipulate it themselves. For instance, databases often have information on the gender, social class, age and ethnicity of respondents, so further statistical analysis can be undertaken to see if these variables are correlated with GCSE grades.

Researcher insights

I (Sarah) published a piece of research with colleagues (Koch et al. 2020) into the question of how social polarisation (increasing inequality) was experienced in four towns in the UK. We used official statistics to examine the income, class, ethnicity, etc. of the towns' inhabitants. The towns were deliberately chosen to permit comparison. All had similar population sizes and were situated within a similar distance of a large city. Two towns were relatively prosperous and two were associated with deprivation.

The data gave important insights into the ways that all four towns were internally polarised but with differing effects. In the wealthier towns, 'elite-based' polarisation meant that poorer communities were located on the edges of the town. In the two less wealthy towns, 'poverty based' polarisation saw poorer communities concentrated in the town centres. In all four towns, the population was divided into an elite (economic, cultural and creative) and a poorer, working-class constituency. There were no middle classes.

The statistical data revealed the geographical effects of polarisation, but it could not tell us about what it was like to live in these towns and so qualitative data was also gathered. Moreover, to enable comparison of the towns we had to rely on data from larger local authority areas rather than the smaller electoral wards of which the towns are comprised. For example, in Margate, we used town-level data rather than data from two key electoral wards, Cliftonville West and Margate Central – two of the most deprived wards in the UK. While the town data gave good insights, it could not capture the extent of extreme deprivation in some pockets of the town. If the team had been able to focus down to ward-level data, we would have achieved a more detailed assessment of the differences between the towns.

To assess income levels, we also used Pay as You Earn income (labour income paid by an employer) and cash benefits (such as child benefit, unemployment benefit and disability benefits), provided by the ONS. However, this data did not include income from self-employment, property income or other investments. As a result, we may not have fully captured divisions in wealth.

How rigorous are official statistics?

Rigour refers to the robustness and usefulness of the research evidence. Statistical data can help us answer important sociological questions. However, there are two questions that sociologists must ask about existing datasets to assess their value.

Is the data sufficient?

While the identification of patterns within statistics is useful, it does not always sufficiently explain any statistical correlations. For example, statistical data might allow researchers to make correlations between different variables – for example, social class and GCSE results – but these findings alone do not allow us to explain why the relationship exists.

Is the data reliable and valid?

Official statistics use categories that require sociological scrutiny. For example, how is social class measured in the dataset? This means that sociologists sometimes question whether official data is valid.

Official statistics are driven by the interests of government, not researchers. For example, sociologists are interested in finding out about the relationship between social class and life chances, but governments do not systematically collect this information. This can affect validity and reliability.

Governments also have political agendas. For example, they might want to find out how many people are committing knife crime, so they encourage police forces to concentrate on documenting this crime. As a result, the crime rates might appear to spike. Sociologists would ask whether this is an actual spike in the rates, an outlier or a result of the way the data is being collected (referred to as a **statistical artefact**).

Official statistics do not always capture complete and accurate data. For example, victims of rape often do not feel comfortable telling the police about their experience, so the crime is underreported. This is what sociologists call the **statistical iceberg**. Moreover, police may investigate certain types of crime more than others. For instance, research suggests that police are more likely to investigate benefit fraud than tax evasion. Therefore, statistical data will not always represent a true picture of criminal activity.

Which sociologists are likely to use official statistics?

Positivist sociologists often use existing statistical databases. They assume that the statistics are a valid representation of reality, revealing facts about the social world and enabling sociologists to undertake objective analysis. For example, a positivist researcher interested in mental illness might use government statistics to show that more anxiety and depression is recorded among young people now than in the past. From this, they might hypothesise that the phenomenon is a result of living in a more stressful and individualised society, in which experiences are dominated by social media.

Interpretivist and constructionist sociologists, who are interested in the way that people construct the social world, suggest that statistics are a social construction and reflect dominant assumptions about the world rather than representing objective facts. Sociologists in this camp would not use the statistics to explain a phenomenon, but they might consider why statistics take the form that they do. Using the same example as above, interpretivist and constructionist sociologists might ask: Are young people now more likely to disclose mental vulnerability because there is less **stigma**? Are doctors more likely to diagnose general 'life problems' as mental health issues? Rather than viewing mental ill health as increasing, sociologists can use statistics to ask deeper questions about the *meaning* of mental health today.

Realist sociologists, such as some feminists, Marxists and critical race theorists, would ask questions about the power relations that sit behind the statistics. They would not take the statistics at face value, but would instead use their theoretical lenses to reveal patterns of inequality in the way that statistics are collected. For example, they might want to show that working-class people, women and some minoritised ethnic groups are more likely to be diagnosed with mental illness. Rather than revealing real differences in the rates of mental illness within these groups, official statistics can be biased and reflect underlying prejudices that mean certain groups come to the attention of the medical profession more than others.

Key concepts

statistical artefact
an inference that results from the way that data has been collected or manipulated; as such, the data may be a consequence of measurement error rather than a reflection of the real world

statistical iceberg
a situation in which official data does not record all phenomena being studied (e.g. crime) and in which there may be many more instances of a phenomena that remain invisible; official statistics often reveal only the tip of the iceberg

Activity

Make a list of all the questions that sociologists must ask about official statistics when using them as data.

Why do some sociologists take official statistics at face value, while others question their validity and reliability?

CLASSIC RESEARCH: SUICIDE

Durkheim, E. (1897/2005). *Suicide: A study in Sociology*. London: Routledge.

Study overview

The most famous example of sociological research that uses official statistics is Durkheim's study of suicide. He was not interested in the individual motivations that sit behind the decision to end life; instead, he wanted to explore what wider social factors were correlated with suicide rates.

Method and sample

At the end of the nineteenth century, when Durkheim undertook his research, the common theories about suicide referred to a country's climate (weather) and psychological factors. Durkheim argued that neither of these factors predicted or explained the differential suicide rates that he observed across Europe. His research explored the complex relationships between suicide and other **variables**, such as:

- Religion: Do suicide rates differ among Protestants and Catholics?
- Nationality: Are suicide rates higher in some countries? To assess this, Durkheim examined data from Germany, Prussia, Switzerland, Norway and Sweden, as well as some information about France.
- Occupation: Do suicide rates differ between professions?
- Marriage: Is death by suicide more common among single and widowed people?

Durkheim also looked at the data longitudinally, sampling data from 12 different time periods. This showed that suicide rates dropped during wartime and increased during economic recessions.

He used **multivariate analysis** to manipulate the official data. That is, he took one variable – religion – and compared it to suicide and then progressively added in the other variables to see if the relationship was stronger, weaker or remained the same.

Findings

Durkheim tested different relationships between variables and suicide rates. He worked consistently through each variable to eliminate it from his enquiry. For instance, he showed that in Germany, Protestantism was a stronger predictor of death by suicide than nationality.

Having shown a strong relationship between religion and suicide, Durkheim moved on to explain why this was the case, by developing hypotheses. For example, he noted that Protestants value individualism, and he suggested that these traits made people more vulnerable to suicide. In other words, he hypothesised that people living in communities without strong shared norms are more likely to take their own lives. He referred to this as 'egoistic suicide'. As his analysis progressed, he identified three further types of suicide:

- altruistic suicide, which occurs in groups with lower levels of individualism and very high levels of social integration (such as suicide bombers)

> **Key concepts**
>
> **variable** a factor measured in sociological research; variables can be dependent (i.e. they change) or independent (be seen to have an impact/effect on other variables) – e.g. social class (independent) has an impact on educational attainment (dependent)
>
> **multivariate analysis** an analysis that involves two or more variables – e.g. a study that considers the impact of social class and gender on educational attainment

- anomic suicide, which occurs when there is rapid social change and when there are lower levels of social regulation, meaning people may experience high levels of stress without any support structure
- fatalistic suicide, when an individual feels that they are subject to too many rules and too much regulation (such as prisoners).

All of these suicide types could be linked to the extent to which people were integrated into, and/or regulated by society. Where integration or regulation was too high or too low, death by suicide was more likely to occur.

Evaluation

Durkheim thus engaged in the impressive testing of the factors that predict suicide. However, he did not assess all variables. For instance, he did not look at the relationship between suicide and living in cities or rural spaces.

Some sociologists have criticised Durkheim for accepting the statistical suicide rates as valid. For example, Atkinson (1973) argues that suicide statistics are a social construction. He suggests that coroners in Catholic-dominated societies might be less likely to record a death as suicide because of the stigma it might bring to the family.

You can evaluate Durkheim's study using the STRIVE criteria as follows.

- Sociological: Durkheim's work makes clear connections between personal troubles and broader social structures. Suicide has been (and continues to be) treated as a psychological problem of an individual. However, Durkheim's work clearly challenges this way of thinking, revealing suicide to be socially caused.
- Theoretical: Durkheim was a positivist, keen to develop objective methods similar to those used in the natural sciences. As such, his work is blinkered to the influence that emotions and identity may have on someone's decision to take their own life. Durkheim also did not consider the realist questions of how oppression and inequality might impact suicide, or the constructionist question of how understandings of what counts as suicide are shaped by social values and beliefs.
- Representative: Durkheim's work is considered representative, as he used official datasets.
- Impactful: Durkheim's study had a profound effect on how we understand suicide. It also established the distinctive value that sociological analyses afford.
- Valid and reliable: Durkheim was confident that the data was both reliable and valid. However, this is open to criticism. Validity assumes that the official data represents the actual number of suicides, but because suicide was stigmatised at the time Durkheim was researching, it is likely that some deaths by suicide were not recorded as such. In addition, each country captures data differently, which raises questions of reliability and whether it is possible to compare the data.
- Ethical and reflexive: Suicide is a highly emotive and difficult area for sociologists to study. By examining statistics, Durkheim avoided the ethical dilemmas that interviews with the families of the deceased would evoke. By studying statistics, he was able to take a detached position, setting aside his own views and values.

CONTEMPORARY RESEARCH: INEQUALITY: THE CLASS CEILING

Friedman, S. and Laurison, D. (2020). *The Class Ceiling: Why it Pays to Be Privileged.* Bristol: Policy Press.

Study overview

Sociologists use the term 'glass ceiling' to refer to invisible barriers that prevent certain people – often women – from progressing at work. In 2020, Friedman and Laurison published a book in which they suggested that barriers were also preventing working-class people from progressing in life. They called this the 'class ceiling' and they used official statistics to explore how it operated. Specifically, they examined whether working-class people were being prevented from succeeding, and whether certain advantages enabled middle-class people to reach the top of their careers and earning potential.

The researchers were interested in testing whether there was a **meritocracy** in the UK. Do all people who are talented and work hard experience upward **social mobility** or are some people being helped up the ladder?

Method and sample

While Friedman and Laurison used some interviews, we concentrate here on their analysis of statistics collected using the Labour Force Survey (LFS). The LFS is the UK's largest employment survey, conducted by the ONS.

The LFS enabled the research team to access a sample of 108,000 respondents, who all lived in private households. The researchers regarded this to be a nationally representative sample. Indeed, they described the survey as the 'gold standard' (p. 240). Their analysis compared each respondent's occupation with their educational attainment and parental occupation, as well as with their gender, ethnicity and disability.

The research was limited to 2013–16, as these were the only years when a new question about parental occupation was included. From 2013, respondents were asked about the job of the main income-earning parent when they were 14 years old, which allowed comparisons to be made between parental occupation and the respondent's own occupation. As such, the researchers could not look at changes across other time periods.

Similarly, while the LFS records people's current occupation, it does not have a quota to fill to ensure that all types of occupations are included in the survey. Moreover, because the LFS focuses on private households, people living in communal accommodation, such as army barracks, were not included in the survey.

To measure people's social mobility, the researchers compared occupational origin (a parent's occupation) and occupational destination (their own current occupation). They thought this would allow them to test whether children from

> **Activity**
>
> Note down any advantages and disadvantages of using the LFS database that you can think of.

more advantaged backgrounds were more likely to progress in careers than those from lower social class family backgrounds.

Findings

By analysing this data, the researchers were able to classify their respondents according to their occupational origins. They used three broad groupings:

- professional and managerial
- intermediate
- working class.

Social researchers must consider certain factors when using the job of the main income-earning parent as an indicator for occupational origin:

- the extent to which households have assets and wealth
- the fact that parental income may be unequal – for example, one household may have only one earner, another may have one high earner and one low earner, and another might have two very high earners
- the assumption that income is related to social class and that income alone explains advantage or disadvantage, when in reality low-earning households may give their children many advantages, particularly if they emphasise education and cultural activities.

To assess occupational destination, the researchers wanted to identify who was most likely to enter what they called 'elite' occupations. The table on the following page shows these occupations and the numbers of respondents in each one.

> **Questions**
>
> Do you think using the job of the main income-earning parent is a valid indicator of occupational origin? Why, or why not?
>
> What problems might there be in using occupation as an indicator of class?
>
> What other indicators/ proxy measures of occupational class origin could be used?

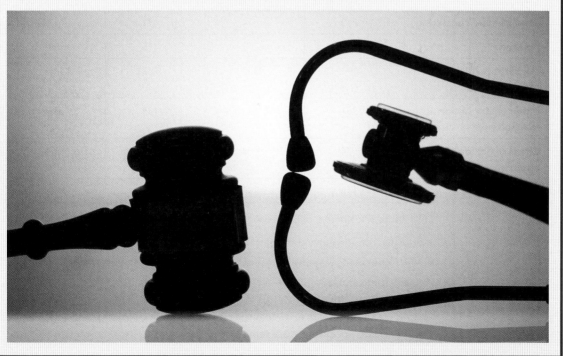

3 OFFICIAL STATISTICS

Occupation	Professional and managerial origins	Intermediate origins	Working-class origins	Total
Performing arts	184	123	74	381
Film and TV	185	139	43	367
Journalism	244	107	39	390
Architecture	92	70	16	178
Academia	298	137	87	522
Science	269	166	81	516
Life sciences	269	123	46	438
Medicine	520	146	47	713
Law	313	124	66	503
Accountancy	384	272	155	811
Engineering	451	401	250	1102
IT	1027	717	433	2177
Advertising	612	311	203	1126
CEOs	103	62	31	196
Management consulting	240	114	71	425
Finance	321	245	115	681
Corporate senior management	494	1152	767	3413
Public sector senior management	281	212	133	626
Chiefs of fire and ambulance services	74	63	42	179
Any other top job	1542	1256	871	3669

Table 3.1: Number of LFS respondents in each elite occupation by origin.

When considering how we classify elite occupations, bear in mind that:
- the categorisation assumes that 'elite' equates with 'high-earning'
- some professions, such as the clergy, are not in this list but might be seen as having considerable power and status that would classify them as 'elite'
- some of the elite occupations on the list do not necessarily command high wages (for example, academia).

Based on their analysis of the data, Friedman and Laurison suggested that we live in a **closed society** (as opposed to an **open society**). They made two important findings:

- People born into families with parents in elite occupations tend to secure elite jobs. To put it bluntly, those at the top stay at the top.
- People at the bottom of the social hierarchy find it more difficult to secure high status jobs even when they have excellent qualifications.

The researchers also highlighted the hidden structures that prevent working-class people from being more upwardly mobile. These hidden structures include: the financial support given to those of middle-class origin; the fact that middle-class people felt more comfortable interacting with elite peers; social networks, which helped people with middle-class origins into good work placements; the fact that working-class people often experienced imposter syndrome (a feeling of not fitting in) and elite people often shut working-class people out.

Overall, the data suggests that a class ceiling exists in the UK, which is holding back capable working-class people. The researchers found that:

- overall, half of the people in the highest-ranking (elite) occupations had parents who had undertaken similar work
- only 20% of people working in elite occupations had a working-class background
- some elite jobs were more closed than others – for example, in medicine almost 80% of people were from privileged backgrounds, and almost a quarter of children of doctors became doctors themselves.

Friedman and Laurison interrogated the data to find out if working-class respondents with the same level of talent were prevented from entering elite jobs. They were able to show that the number of working-class people who went to high-ranking universities and achieved first class degrees was higher than the number of working-class people in high-ranking jobs. They inferred from this that the playing field is not level: 'Even when working class students outperform the privileged, they are less likely to get top jobs' (p. 39). The research showed that while 'Nearly two-thirds (64%) of privileged-origin respondents who achieved first class degrees from Russell Group universities progress to a top job, less than half (45%) of those from working class origins with exact same achievements do so' (p. 38).

Two other key findings from this analysis suggest that social class origin intersects with other socio-demographic characteristics to predict social mobility. First, women, many people from minoritised ethnic communities (especially those of Black, Bangladeshi or Pakistani origin), and those with disabilities were also underrepresented in the top jobs. Second, while women earned less than men in the same job, the differences were more profound if the women also came from working-class backgrounds. 'Working class women, on average, earn £7,500 pounds less than privileged-origin women, who in turn earned, on average, £11,500 less than privileged origin men' (p. 50).

> **Activity**
>
> What data can you extract from Table 3.1 to support the researcher's claims?
>
> Is there anything that refutes these claims? If so, what?
>
> What other questions might a sociologist ask about the data?

> **Question**
>
> While the results show a correlation between class origin and class destination, you could still attribute this correlation to merit – to the talent and hard work of children born into elite families. What additional data would the researchers need to gather to corroborate or question whether meritocracy exists?

The researchers used these findings to suggest how companies could be more meritocratic. For instance, they suggested that unpaid and unadvertised internships should be banned.

Evaluation

This study tackles an important and enduring sociological question – that of inequality and social mobility – and identifies the myth of meritocracy at work. The research is also intersectional in that it highlights how people of different genders, social class and ethnicities are differentially impacted by the class ceiling. The research aimed to be impactful by raising awareness of the class ceiling and advising companies on how to enhance rather than limit social mobility. The study takes a realist position in that it assumes that social inequalities are real, that there are hidden structures that prevent social mobility, that these hidden structures can be revealed, and that they can be challenged.

Using the Labour Force Survey (LFS) was useful, as it enabled the researchers to access a large pre-existing dataset. Because of the size of the LFS, the sample is representative. It had also been trialled and tested and so was likely to be reliable. However, as with all research that uses secondary data, the researchers would have been limited to the operationalisation of concepts used within this dataset (i.e. how concepts like social class were measured by the LFS). In this regard, it is hard to say with certainty that the results are valid (for example, their categorisation of 'elite' occupations and their classification of class origin could only be determined by the data that was readily accessible to them in the existing dataset). The study does not raise any ethical issues, as it uses official statistics.

CONTEMPORARY RESEARCH: EDUCATION: THE GLASS FLOOR

McKnight, A. (2015). *Downward mobility, opportunity hoarding and the 'glass floor'*. Research report. Social Mobility and Child Poverty Commission.

Study overview

Where Friedman and Laurison (2020) identified a 'class ceiling' that stopped working-class individuals moving up the social ladder, McKnight was interested in the factors that created a 'glass floor', which prevented middle-class people from experiencing downward social mobility.

Method and sample

McKnight tracked children's educational performance and life experiences over time (longitudinal analysis), using the British Cohort Study. This study follows around 17 000 people who were born in a particular week in 1970. It has gathered huge amounts of data, using methods such as interviews, standardised tests, health screenings, parental questionnaires until the cohort was 16, and questionnaires given to the cohort after they reached 16. The data has been gathered at distinct stages of life, with specific check-in points at ages 5, 10, 16, 26, 30, 34, 38 and 42.

While the BCS focused on all births in a single week in 1970, there were some inconsistencies in data sampling, collection and analysis, which had implications for McKnight's study:

- After the age of 16 it became harder to track participants because they were no longer in school. The response rate went down over time and therefore **attrition** was an issue for the study.
- At the check-in points, it became clear that the sample of children born in the same week had grown to include people who had migrated to the UK. These children were added to the BCS sample to make the dataset more representative over time. However, McKnight excluded these children from her study because she wanted to compare data from every check-in point.
- Some of the check-in points were not equally spaced as they were initially intended to be, due to unforeseen challenges with data collection. For instance, the original idea was to collect data at the age of 15, but strike action by teachers who were responsible for administering the tests meant that the survey was delayed and was also incomplete.
- Responsibility for the survey was taken over at different points by different organisations due to issues of funding.

McKnight based her research on the children in the initial sample who could still be contacted at the age of 42. She excluded the 766 people who had joined the study later. In total, McKnight's research tracked 9075 people – 57% of the original sample.

McKnight's research objective was to examine children of different abilities and backgrounds (measured at the ages of 5, 10 and 16) and compare these starting points to their adult outcomes at 42, to measure the extent of their social mobility. McKnight chose this age, rather than 34 for example, because she believed that at this point, the cohort was mature enough to measure their adult outcomes (income and occupation) with confidence.

To classify the cohort into different starting positions, children at the age of five were grouped according to their performance in cognitive tests and the social position of their family. At five years old, the children were given five different educational tests that enabled researchers to split them into groups of high and low attainment. The researchers were confident that this was a reliable measure because they used the average score from the five tests rather than relying on one test alone, which could be skewed by luck or poor performance on the day.

To assess social position, McKnight used both the income and occupation of the family to decide if the children came from advantaged or disadvantaged backgrounds. She took a gender-neutral position when determining the 'family' occupation, choosing the highest-ranking occupation, whether that was held by the mother or father. Nevertheless, only one occupation was recorded. McKnight

> **Questions**
>
> What do you notice about these different time points? Why do you think the researchers chose these check-in ages?

Key concepts

attrition a decline in the number of research respondents over time

Questions

What problems might there be with excluding people from a study and working with a smaller sample, as McKnight did?

From what you know about sociology, how useful are tests for measuring ability? What might challenge the reliability of these tests?

What do you think about the labels McKnight used: 'advantaged' and 'disadvantaged'?

Activity

Rank the four groups in terms of who you think is more likely to be successful as an adult.

What made you decide to put them in this order?

What hypothesis could you develop to support this ranking?

---

used the occupational category of the highest-earning parent to create two broad categories of 'advantaged' and 'disadvantaged', and split the children into each of these groups.

Findings

Together, the two measures of attainment and social position produced four main groups that could be tracked to see how starting categories impacted adult success. In other words, McKnight was able to undertake a form of survey experiment, by comparing the different groups over time. While such binary categories enable comparisons, they also hide the complex differences that exist within each group.

	Advantaged social-class background	Disadvantaged social-class background
Low attainer	A	B
High attainer	C	D

Table 3.2: The four main groups divided by social background and attainment.

As you might expect, Group C was the most successful in later life. The combination of educational talent and social background predicted success. Group B was the least successful in later life.

However, the position of Groups A and D was not so straightforward. While you might expect group D to be ranked ahead of Group A, this was not supported by the statistics. The research found that in both high and low attainment groups, children from families with higher incomes or more advantaged social backgrounds had a greater chance of achieving professional success than those from less advantaged groups. The researchers showed that children from advantaged backgrounds with lower educational scores avoided downward social mobility.

To establish why this happened, the researchers looked at the cognitive tests taken at ages 10 and 16. These revealed that advantaged children from the low-attainment group had shown improvements in educational ability, due to factors such as intensive tutoring, attending grammar and private schools, and receiving support from their parents, who were more likely to have a university education.

In short, McKnight argued that advantaged children are given lots of extra educational support to improve their social position and therefore they overtake able but disadvantaged children, hoarding opportunities for success (**opportunity hoarding**). For instance, McKnight found that less able but more advantaged children had higher self-esteem and were also more likely to go on to university by the age of 26 compared to their able but disadvantaged peers.

In contrast, high-attaining children from disadvantaged family backgrounds were less likely to be able to convert this talent into success in work in their adult lives. In other words, despite talent, the cleverer children from this social position did not achieve the same income and types of occupation as their advantaged peers.

This research corroborated the idea of a 'glass ceiling' that stopped able but disadvantaged children from progressing. It also revealed the existence of a 'glass floor'. Opportunity hoarding prevented less able but advantaged children from falling (downward social mobility) and meant that they did better in life than more able children who came from poorer families.

The findings from the research revealed the advantages of attending selective and private schools and of receiving help from tutors and family members.

The suggestion is that for meritocracy to work, clever but disadvantaged children need more support from schools to offset the help being given to their advantaged peers. One practical suggestion made by McKnight was that the best teachers should be incentivised to work in schools that disadvantaged students attend.

Questions

McKnight's research reveals what she calls 'opportunity hoarding'. Other sociologists refer to similar concepts such as **cultural capital**. Consider other sociological studies that you already know about. In what other ways do middle-class parents give their children advantages? What are the impacts of this advantage?

Activity

Compare McKnight's study with Friedman and Laurison's study of the class ceiling. Which study do you think had the most rigorous methods and why?

Evaluation

McKnight's study clearly addresses a core sociological concern about educational inequality. As with Friedman and Laurison's study (2020), McKnight's use of official statistics enabled her to prove a relationship between social advantage (class) and educational success.

However, the study also highlights some of the problems with using official statistics. For example, McKnight had to rely on pre-existing questions and concepts. In addition, her sample size was smaller than she might have liked and was not representative of all ethnicities.

Activity

Write a list of all the problems you can think of associated with using official data.

Case study: Official statistics in education: the achievement gap

Sociologists are interested in using official statistics to track and understand differential educational achievement by social groups. However, they also express caution about some of the inferences that can be drawn from statistics. For example, when researching attainment by ethnicity, the use of wide-ranging descriptors such as British Asian or Black British, can mask the diversity (heterogeneity) within these communities. More than this, focusing on overall (homogenous) statistics has served to shape research interests and has meant that much research has examined underachievement rather than considering where achievement is higher than average within minority ethnic communities. Additionally, researchers have tended to focus on the *achievement* of ethnic minority students rather than looking at the institutional factors that can explain why students are *awarded* different grades. For example, researchers have tended to blame family values and dynamics for lower educational outcomes rather than look at how institutions, curriculum and assessment types favour some communities over others.

Overall, official statistics suggest differential educational outcomes by ethnicity, as Demie (2021) summarised:

> In 2018 it was identified that amongst those ending their compulsory education in the UK, Black Caribbean and Pakistani pupils were least successful academically, with only 44% of Black Caribbean and 50% of Pakistani pupils achieving 9–4 grade GCSEs in English and Maths. In contrast, around 77% of Chinese, 86% of Indian, 61% Black African and 60% of White British pupils achieved above the national average. Bangladeshi, Indian, and Chinese pupils also achieved better than African heritage pupils. (p. 3)

However, there is research evidence that questions this overall picture. Strand's research (2012; 2015), for instance, has shown that pupils from most minority ethnic groups make good progress during secondary school and exhibit greater resilience to deprivation relative to their deprived White British peers. However, Strand has also shown that Black British Caribbean and Black British African pupils from more advantaged homes experience underachievement when compared to their White British peers. This research has drawn attention to the impact that in-school factors, including racism, rather than familial factors, have on the low attainment and poor progress of Black British Caribbean pupils.

Gillborn et al. (2017) draw on critical race theory to analyse official statistics and question the way that measures of educational attainment are decided by government. They show that government statistics consistently suggest that there is an achievement gap for British Black Caribbean students, and this has historically been used to blame ethnic communities rather than broader educational prejudice and discrimination.

To counter this prevailing underachievement narrative, the researchers examined 25 years of data from the Youth Cohort Survey and the National Pupil Database. Rather than take the data at face value, they looked at the policy changes that have had an impact on the ways that attainment has been variously operationalised over time. They showed that changing ways of measuring attainment widens the gap between Black British and White British performance. For example, when attainment was measured by five or more GCSEs in any subject area, the attainment gap narrowed over time. However, the introduction of the English Baccalaureate (EBacc) in 2010, which meant that English, Maths, Science, History or Geography was mandated as the performance measure for schools, saw the previously closing attainment gap between ethnicities reappear.

> The data suggest that the introduction of new 'tougher' benchmarks not only reduces the overall scale of 'academic success', it also sets back progress toward race equality by restoring historic rates of disadvantage. (p. 865)

The consequence of this was that 'In England to date, such changes have had a marked regressive and racist impact; redefining the benchmark has led to a wider Black/White gap' (p. 868).

Case study: Official statistics in education: Debunking myths

In 2018, Khattab and Modood explored British Muslim educational attainment using the Labour Force Survey and the National Pupil Database, undertaking a more **heterogeneous** examination of this broad ethnic category. As they note, most research to date has focused on Pakistani, Bangladeshi and Indian communities, which make up about 60% of the British Muslim population, and this has skewed our understanding. Instead, as well as Pakistani and Bangladeshi students, their study included White, Black, Arab and mixed heritage Muslim students.

Using the existing data, the researchers assessed the independent impact that gender, religion, family composition, social class (measured by occupation), cultural capital (measured by activities such as learning a musical instrument), parental and student expectations, and student effort all had on educational outcomes.

They found the following:

- Muslim students outperform at GCSE level, and beyond this they seem to be performing as well as the non-Muslim population, including attending Russell Group universities. As the researchers show, this finding unsettles the usual portrayal of Muslim students as educationally disadvantaged compared to when researchers just focus on British Pakistani and Bangladeshi students. The way that official statistics are used can change understandings of educational disadvantage and shape public and policy narratives.
- Muslim girls are outperforming Muslim boys, especially at school – a finding that contradicts many cultural assumptions.
- There are high levels of child and parental educational expectation, which again challenges many educational stereotypes.

Shain's qualitative research corroborates these findings (see Chapter 4).

CONTEMPORARY RESEARCH: CRIME: POLICE 'STOP AND SEARCH'

Bowling, B. and Phillips, C. (2007). 'Disproportionate and Discriminatory: Reviewing the Evidence on Police Stop and Search'. *Modern Law Review*, 70(6), pp. 936–961.

Study overview

Bowling and Phillips (2007) sought to reveal the hidden power structures that shape the way that official statistics are presented. To do this, they examined the statistics that capture police powers to stop and search individuals in public spaces. Several government acts (for example, Section 1 of the Police and Criminal Evidence Act) give the police permission to stop individuals and search them, based on suspicion or on intelligence they have received about a particular area or event, rather than on visible criminal activity. These powers have been criticised, with claims that the police racially discriminate by disproportionately targeting Black British communities and those from minority ethnic groups for stop and searches. This is referred to as institutional racism.

The researchers wanted to investigate two things. First, they wanted to examine whether stop and search rates varied disproportionately by ethnicity. Second, they wanted to examine whether any disproportionate statistics were a true measure of crime, or whether they could be explained by other factors, such as racial discrimination.

Becoming a social researcher

Bowling and Phillips published their study in 2007, but you can also look at data from 2020 and examine stop and search rates by ethnicity, which show differences between ethnic communities to be an enduring finding (Figure 3.1).

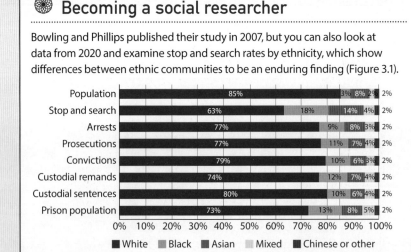

Figure 3.1: Ethnicity proportions for adults throughout the criminal justice system, 2020. (percentages have been rounded up or down to the nearest percentage point)

Look at the relationship between ethnicity and indicators of criminality. You will notice that for all areas of criminal activity, White people are the largest group of offenders. However, the population tab at the top of the chart reveals the following:

- The White population is underrepresented in every category of the Criminal Justice System (CJS).
- The Black population is disproportionately represented in these statistics; Black communities are more likely to enter the CJS at all levels, and particularly in stop and search (six times more likely than would be expected from Black population size).
- The Asian population is disproportionately stopped and search but otherwise is proportionately represented throughout the CJS.
- The mixed population is disproportionately represented throughout all levels of the CJS.
- The Chinese population is almost the same throughout.

Question

Official statistics do not always include the population size bar. How would the exclusion of the population size bar change your interpretation of the figures?

Method and sample

Various sets of official statistics capture data on stop and search. Since April 1996 the Home Office has required the police to monitor the ethnic origin of suspects and it has produced annual reports of statistics on 'race' and the criminal justice system. Bowling and Phillips looked at the relationships between ethnicity and official stop and search data.

Findings

Bowling and Phillips found that people from Black British communities were six times more likely to be stopped and searched than people from White British communities in in England and Wales – evidence of disproportionate use of stop and search.

The researchers argue that a set of empirically verifiable facts can be seen in the statistical data. That is, Black British and minority ethnic groups are more likely to be stopped and searched than people from White British communities. However, it is important to ask whether the collection of this data is methodologically sound.

The researchers describe some conceptual and methodological problems with the collection of data on ethnicity within the CJS:

- Until April 2005, it was the police officer who recorded the ethnic origin of the person stopped.
- Recording rates may vary. Not all stop and searches are recorded by the police and police officers are more likely to record their stops of Black British people than White British people.
- Recording instances of stop and search varies across police forces geographically.

Questions

Why might police recording of ethnic origin be a problem when capturing official statistics?

When researching crime, sociologists can draw on police-recorded crime statistics or the Crime Survey for England and Wales. What are the advantages and disadvantages of each?

Research tip

You can access government crime statistics at: https://www.crimesurvey.co.uk/en/index.html

Questions

What theoretical position do you think sits behind this research project?

Based on their study, what are some of the disadvantages of using official statistics?

Key concepts

statistical discrimination when statistical patterns are taken at face value and reinforce discriminatory practices; the concept emerged from research into ethnic differences in the labour market (Phelps, 1972), which showed that employers believed productivity levels differed between ethnic groups (though not necessarily for reasons of ethnic difference) and would use ethnicity as a predictor of a potential job candidate's abilities

Nevertheless, while there are methodological problems, Bowling and Phillips suggest that the police records are a valid and reliable indicator of disproportionate use of powers with different ethnic groups. By comparing data from other surveys, including the British Crime Survey (now called the Crime Survey for England and Wales), they found similarly disproportionate rates of stop and search by ethnicity.

Next, Bowling and Phillips wanted to establish whether these disproportionate rates were in fact discriminatory.

The police argue that they target particular geographical areas to implement stop and search based on crime rates in those areas. Waddington et al. (2004) argue that it is coincidental that these areas tend to be populated by Black and minority ethnic groups. These groups simply make up the majority of the available population that might be stopped and searched. They conclude that the disproportionate rates are not discriminatory but are a valid representation of crime in localised areas.

Bowling and Philips take issue with this explanation. They argue that police being guided by crime statistics to determine where they concentrate their activity presupposes that crime statistics are themselves accurate. However, police discrimination can also explain the higher crime statistics in particular areas. Bowling and Phillips argue that this is a form of **statistical discrimination**.

Evaluation

Bowling and Phillips's research shows from a social constructionist perspective that official statistics can be misleading and discriminatory. The way that statistics are presented can influence how we interpret them. Statistics can also be used to target people unfairly and disproportionately, in this case on the basis of ethnicity.

CONTEMPORARY RESEARCH: FAMILY: DOMESTIC VIOLENCE

Walby, S., Towers, J. and Francis, F. (2014). 'The decline in the rate of domestic violence has stopped: Removing the cap on repeat victimisation reveals more violence.' Violence & Society Research Briefing.

Study overview

In this research briefing, Walby et al. examine official statistics on rates of domestic violence from the Crime Survey for England and Wales (CSEW) rather than police-reported crimes. This source is deemed to be more reliable because victims of domestic violence may not always report crimes to the police. Nevertheless, the study wanted to show that the statistics from victim crime surveys can also be analysed and presented in problematic ways.

 Thinking further

In 2020, the police recorded a total of 1 288 018 domestic abuse incidents and crimes. In contrast, the CSEW estimated 2 300 000 incidences of domestic abuse. Consider why figures on domestic violence as recorded in the CSEW, which asks victims about their experiences of crime, might be more reliable than police-recorded crime. There are many explanations, including:

- Women do not report domestic violence to the police.
- The police record domestic violence inconsistently, because domestic violence does not have its own 'crime code'.
- The police tend to record gender-based violence under other codes of homicide, rape and sexual offences.

Method and sample

The CSEW draws on a sample of around 40 000 people per year and is largely considered to be representative. It is regarded as sociologically important because it includes crimes that are not always reported to or recorded by the police. It also collects socio-demographic data on the respondents.

Questions about domestic violence in the CSEW are asked in two ways, through a face-to-face and a self-completed section of the survey. There is an interesting methodological observation to be made here: women are 3.8 times more likely to disclose domestic-violence incidences in the self-completed section of the survey than they are face to face.

Questions

Why might surveys provide more accurate accounts of whether women have been subject to domestic violence than interviews?

The CSEW only asks questions about crimes that have taken place in the 12 months before the interview. What challenges might this create for researchers wanting to understand rates of domestic violence?

Findings

Walby et al.'s research revealed that at the time of their study (2014), the declining rates of domestic violence since the mid-1990s had stalled (stopped falling), while the rates of other violent crimes had continued to fall.

CSEW data only records up to five crimes in the official statistical data. This means that even if the victim has reported six or more incidences, the official statistics still record it as five. The ONS (2013) defends this capping of crime reports because it believes that only a very small number of people experience more than five crimes. However, when this cap was removed, 60% more violent crimes were recorded for men and women.

Look at Table 3.3, which comes from Walby et al.'s research and which breaks the data down by sex. This data showed that when the cap was removed there was a 70% increase in the number of violent crimes against women (compared to a 50% increase in the number of violent crimes against men)

> **Question**
>
> Why might the capping of the number of crimes be an issue for recording cases of domestic violence?

	Females		Males	
	Estimated number of offences 'capped'	Estimated number of offences 'uncapped'	Estimated number of offences 'capped'	Estimated number of offences 'uncapped'
Domestic	246 000	419 000	103 000	170 000
Acquaintance	368 000	760 000	435 000	801 000
Stranger	225 000	238 000	606 000	782 000
Total	**839 000**	**1 417 000**	**1 144 000**	**1 753 000**

Table 3.3: Estimated numbers of violent crimes (violence against the person and sexual offences) by domestic, acquaintance or stranger, by sex of victim, CSEQ, 2011/2, capped and uncapped.

Evaluation

Because Walby et al.'s study uses official data from the CSEW and not police-recorded crime, the data is more likely to be valid and represent true crime figures. However, the researchers also note that the CSEW does not record historical and repeat abuse against women and thus underestimates the extent of domestic violence, which affects the validity of the data. This study shows the importance of scrutinising the methods used to generate official statistical data.

Official statistics as a social research method

1. What are the advantages of using official statistics to research educational attainment and meritocracy? Use examples from this chapter to illustrate your answer.

2. Sociologists doing research using official statistics must accept the existing operationalisation of the concepts that have been used. Drawing on examples from the research studies outlined in this chapter, explain the limitations of this method.

3. Are official statistics social facts or social constructs? Draw on relevant theories and studies to debate this question.

Advantages of official statistics:

- The most obvious advantage of using ready-made statistics is that the time it takes to undertake research is substantially reduced.
- Access is usually free, which means that the research is easier and cheaper to conduct.
- Government statistics are usually based on large samples – certainly bigger than a sociologist or even a team of sociologists could generate. Therefore, the data is more likely to be representative.
- Statistical data can be used to look at patterns longitudinally (e.g. Durkheim looked at suicide rates over time).
- Statistics allow us to answer research questions that could not be answered using other methods. For example, McKnight relied on statistical data to reveal inequalities in social mobility that could not otherwise have been established.
- Official statistics are also often reliable – the measures have been heavily trialled and tested. However, the assumption of reliability is not entirely straightforward, as you can see from Bowling and Phillips' study.

Disadvantages of official statistics:

- Social researchers cannot decide the topics included in government surveys and therefore may not be able to fully answer their own research questions, especially if key variables are missing.
- Some topics are not of interest to government and therefore the datasets are unavailable. For example, while there is a lot of data about how often people visit doctors and hospitals, the government does not collect information about how many people visit complementary therapists.
- It is not always possible to be confident that the data is measuring what a sociologist wants to study – in other words, the validity of official statistics is sometimes questionable. For instance, you might think that measuring causes of death from death certificates would be valid – after all, skilled doctors make the decision and they choose from a set of pre-existing categories, such as heart disease, cancer, influenza or diabetes. Yet think about measuring causes of death during the pandemic. When elderly people were registered as dying from COVID-19, can we be confident that other illnesses (co-morbidities) were not factors? Are sociologists even able to explore this possibility?

CHAPTER 4: SOCIAL SURVEYS

What are social surveys?

Social researchers often want to learn about how the realities of social life are organised and distributed, but sometimes there are no official statistics that can help them. Perhaps they have a new research question and hypothesis but cannot test it using ready-made datasets. For example, a researcher may be interested in how many young people did not undertake their homework during the pandemic, or schoolteachers' attitudes towards the current sociology curriculum. As this quantitative data is not available, sociologists must develop their own surveys.

Developing a social survey, sometimes called a questionnaire, is a practical, cost-efficient and reasonably time-effective way to gather quantitative – and sometimes qualitative – data. Quantitative data is generated using closed questions (a simple tick box or a set of pre-determined options), while qualitative data is extracted by open questions, where respondents have space to write their own answers. By designing their own questions, researchers can ensure they are interrogating their specific research topic and they do not have to rely on pre-existing data that might not be fit for purpose. A bespoke survey also means that the researcher can select their own sample of respondents. In addition, they can choose to go through the questions in person with each respondent (which is more time consuming) or allow respondents to complete the survey at their own convenience, either as a hard copy or online (known as a self-administered or self-completed survey).

Social surveys are a great way to gather lots of numerical data. They can offer insights into population trends as well as the opportunity to test hypotheses. If the questionnaire is well-designed, it is likely to be highly reliable and can be used by more than one researcher.

However, Savage and Burrows (2007) argued that while surveys are an incredibly valuable research tool, they are declining in popularity because surveys produced by governments and other organisations capture data on a much larger scale than social researchers can achieve.

🧹 Social surveys on housework

What can a social survey tell us about work in the home? Oates and McDonald (2006) wanted to understand if men or women were more likely to engage in recycling household waste. They designed a questionnaire and sent it to a quarter of the households in Sheffield that had signed up for a council recycling bin, by choosing every fourth address on the council website. The survey included both open and closed questions to find out who did the recycling.

Of the 1532 questionnaires sent by post, 469 were returned. These tended to over-represent two-person households and under-represented one-person households. The results suggested that women were more likely to initiate and sustain recycling, and that men were more involved if they lived in a household with others rather than alone.

> **Questions**
>
> What type of sampling method did Oates and McDonald use in their research? Look back at Chapter 1 if you need to.
>
> How generalisable is this research?
>
> What questions are left unanswered by using a quantitative method like this?

How do sociologists use social surveys?

Significant work goes into devising a social survey. Many decisions need to be made, ranging from how to measure concepts, what variables and indicators to use, and how to frame the survey questions, to who to sample and how to analyse the data.

Concepts and measurement: Operationalising a research question

All research starts with a question, but sociological questions do not always include concepts that can be easily measured. For this reason, quantitative social surveys must ensure that the complex aspects of social life are broken down into measurable components. This is known as operationalising social concepts, as introduced in Chapter 1. During this process, sociologists use what they call indicators of, or proxy measures for, the concept. For instance, they may be interested in researching the extent to which call centres are alienating places to work. **Alienation** is a key concept in Marxist theory, but how can it be measured in a survey?

Consider the following survey question:

How often do you feel alienated at work?

- All the time
- Most of the time
- Some of the time
- Never

There are problems with this question as it is currently presented. To begin with, the concept of alienation might be understood and interpreted differently by

> **Activity**
>
> Develop a range of social survey questions that would help a researcher understand the extent to which workers in call centres are alienated.

> **Activity**
>
> Make a list of sociological concepts, then choose one or two to operationalise. To start you off, consider how you might operationalise the concepts of cultural capital, poverty and social mobility.

> **Key concepts**
>
> **Likert scale** a scale used to measure attitudes, knowledge, perceptions and values, framed as a series of statements and a choice of options indicating how strongly a respondent agrees with each one
>
> **attitudinal question** a question that measures subjective things, such as opinions, emotions, perceptions and judgments; this type of question almost always uses a rating scale like the Likert scale

respondents (or not understood at all). It is also a very blunt question, which would not reveal much insight into the workplace.

You can see from this example why sociologists seek to operationalise their concepts in meaningful ways. For Marx, alienation was observed when workers could not be creative, when they were in competition with other workers, and when they were only involved in a small part of the work rather than being part of the overall process (such as a call-centre worker who only hears complaints but is not involved in their resolution). Therefore, it is possible to break down alienation into measurable components – for example, by focusing on the extent to which workers feel powerless or isolated, whether they are connected to their fellow workers, whether they feel valued by their employers, and the extent to which they feel connected to the product of their labour (in this case the people they are calling), and so forth. These provide some indicators for the concept of alienation, and from these it is possible to devise questions that can measure it.

Choosing a question type

Questionnaires always contain some closed questions. They sometimes also offer opportunities for respondents to provide more detail in the form of open questions.

With closed questions, respondents choose from a limited range of possible answers. The answers can be quickly collated and analysed, but that does not mean that closed questions are easy to devise. There is a science to the good design of closed questions, and clear, careful wording is essential.

Closed questions come in multiple forms. There might be simple yes and no answers, a list of options, or respondents might be asked to decide how strongly they agree or disagree with a series of statements (not questions), known as a **Likert scale**. A researcher can include the same **attitudinal questions** asked in different ways, to check that the respondents are being consistent in their answers. This ensures the reliability of the answers and the validity of the survey. For example, responses to the two statements in Table 4.1 would be very different.

Statement	Strongly Agree	Agree	Neither agree nor disagree	Disagree	Strongly Disagree
My workplace is free from bullying.					
Bullying is commonplace in my workplace.					

Table 4.1: Examples of attitudinal questions on a Likert scale.

Closed questions can be devised by the researcher, but sometimes examples of relevant questions already exist. For example, when collating information about the socio-demographics of their sample, the researcher can use questions that have already been used in official government statistics (see Chapter 3) to find out about the gender, sexuality, class, ethnicity, disability, etc., of the respondents. These are called **harmonised questions**, and using them enables the researcher to compare their sample with national statistics and to be confident about the consistency of the question wording. There is a whole range of available harmonised questions about educational qualifications, housing, occupation and health status, among other socially important things.

> **Key concepts**
> **harmonised questions**
> questions that have already been developed and trialled by other groups such as governments; these are useful for comparing a sample and responses to other representative data

Advantages and disadvantages of closed questions

Advantages:
- They are easy to complete and have higher response rates.
- Having answers to choose from increases clarity for a respondent.
- They are easy to code and analyse.
- They are reliable.

Disadvantages:
- Fixed choices may not include the answer the respondent wants.
- The researcher does not develop a rapport with the respondent.
- There is no opportunity to check that the respondent has interpreted the question correctly.
- They may lack detail and nuance.
- Validity may be compromised because it is impossible to know how a respondent has interpreted the question and answered accordingly.

Open questions enable respondents to answer as they wish, although there will be some space restrictions. Examples of open questions are:
- *Can you tell us what you enjoy about working in a call centre?*
- *Can you tell us what you dislike about working in a call centre?*

Advantages and disadvantages of open-ended questions

Advantages:
- Respondents are not bound by a particular set of answer choices.
- Unexpected answers may be found.
- They are useful for understanding responses in more detail.
- They have validity because the answers reveal how the respondent has understood the question.

Disadvantages:
- Respondents may not want to answer open questions, as they take more time to complete.
- Respondents may give very different information from one another, which makes the data difficult to analyse.
- Responses may be limited by the respondent's understanding of the question and their literacy in answering it.
- Open questions take time to code and analyse, which makes the process more expensive.
- Researchers may interpret answers differently.
- Open questions are less reliable because the researcher cannot be certain that respondents would answer in the same way at a different point in time.

Wording a question

Great care needs to be taken when creating social survey research questions, whether they are open or closed. Good questions need to be:
- Clearly written: Questions should be unambiguous and should not be open to interpretation.
    - Short questions are always better than long ones.
    - Avoid technical language or concepts.

- Avoid **double-barrelled questions** (more than one question at a time), such as *How satisfied are you with the pay and working hours in the call centre?*
- Statements that include negatives, such as *There is not any bullying at my workplace*, are often misread by respondents so are best avoided.
- Aim for symmetry and balance in closed options. For example, the balance is skewed to the positive in the following question: *Overall, how do you rate working in a call centre?: Excellent, Very Good, Good, Acceptable, Poor.*
- Avoid overlap in answer options so respondents can make a clear choice. For example, if given the age ranges 18–21 and 21–25, there is confusion over which option a 21-year-old should choose.

- **Specific:** Generalised questions are open to interpretation, so always be specific. For example, ask *What did you most enjoy when talking to a new customer?* rather than *What do you enjoy about working in a call centre?* You also need to give clear options in closed questions. For example, the question *How often do you take a break at work – once, twice, three times, more than four times?*, would not reveal how long the breaks were, so it would be difficult to draw conclusions from the same answers.

- **Non-leading:** Avoid **leading questions**, which take respondents in a particular direction. For example, the question *Do you agree that working in a call centre is isolating?* contains an underlying assumption that working in this environment can be lonely.

- **Unblinkered:** Sociologists need to be reflexive when designing their questions, making sure that they do not reproduce common assumptions or entrenched power relations. For example, a question that asks a woman *Does your partner help with the cleaning?* contains an implicit assumption that cleaning is a woman's responsibility because it uses the word 'help'. A better question might be *How much cleaning does your partner undertake?* Similarly, survey questions that only ask about male and female gender categories fail to fully encapsulate and appreciate the diversity of gendered identities. Furthermore, questions might reflect a colonial bias if they make White **normative assumptions**. For example, survey questions that explore why British Bangladeshi boys underachieve in school, but focus only on family support rather than on differential support in school, would stand as an example of methodological racism. This is because the focus of the questions makes an assumption that explanations for educational achievement reside in the family, ignoring structural barriers.

> **Key concepts**
> **double-barrelled question** a question that touches on more than one issue but only allows one answer
> **leading question** a question that prompts a respondent towards a particular answer

Activity

Note down all the advantages you can think of for a) sitting with a respondent and b) asking them to complete a survey on their own. Consider the following:

- Which is more likely to get a better response rate and which might have more missing responses?
- Which is more time-consuming and/or expensive?
- Which is more convenient for the respondent?
- Can you be confident that the right person is actually answering the question?
- Can you be confident that the researcher is not influencing the responses?
- Is there an opportunity to clarify a question if the respondent does not fully understand?
- In which form is it easier to ask sensitive or embarrassing questions?

 Becoming a social researcher

What mistakes can you identify in the following survey questions about housework?

- What age bracket do you fall into? 15–18; 18–25; 25–40; over 40
- What is your social class?
- Do you clean the house?
- How frequently does your husband help with the cleaning and cooking?
- What do you dislike about housework?

Choosing a sample and sample size

Making a choice about your sample and its size can be complicated (see Chapter 1). Researchers using social surveys often aspire to probability sampling methods, as these are more representative. However, this can be challenging, as you will see in the studies below.

Face-to-face or self-completed?

Once researchers have created their questions, they need to decide whether to:

- sit with or call the respondents
- read out the answers
- ask respondents to fill in the survey in their own time.

Undertaking the analysis

Once the survey data has been collected, the analysis can begin. There are several ways to analyse quantitative data generated through social surveys, which involves looking at the relationship between dependent and independent variables. These include univariate, bivariate and multivariate analyses. Analysis of open questions can be qualitative and quantitative. For example, researchers can group the various responses into types and then count how many occurrences of each type there are.

- Univariate analysis is where the answers to one question are collated and compared to give descriptive statistics.
- Bivariate analysis is where a researcher compares the answers to two questions (cross tabulation), which enables them to analyse the relationship between two variables. This type of analysis can suggest correlations between two variables if there is a statistically significant relationship. In Du Bois' study (1899/1996), he examined poverty (variable 1) in relation to ethnicity (variable 2), and showed that not all members of the Black community were poor.
- Multivariate analysis is a statistical technique that enables researchers to look at the relationships between more than two variables. For instance, in his study on suicide (see Chapter 3), Durkheim looked at the relationship between religion (variable 1), nation (variable 2) and suicide (variable 3).

Researcher insights

In 2020 and 2022, I (Sarah) published research that used two separate surveys with teachers of A Level Sociology in England, to find out what sociology teachers thought about the current sociology curriculum. It was difficult to locate a sample frame, as there was no nationally available register of teachers that I could access. In the first survey, with fellow sociologist Anwesa Chatterjee, we called each school and asked for the name and email address of the sociology teachers. We then emailed 1054 questionnaires, from which we received 204 responses. In the second survey, we used Facebook and Twitter to advertise the study; this time the number of teachers who responded doubled (416). As you can see, choosing your communication technique is important.

The research showed that teachers were passionate about sociology and worked hard to ensure that the material they taught was interesting. They reported that students found sociology to be engaging and transformative. However, to validly garner students' views, a specially designed survey for pupils would have been better. The research showed that while sociology was very popular (at the time, it was the sixth most popular A Level in the UK, and the fifth most popular with girls), teachers often struggled to assert the value of the subject with senior management and some parents.

Moreover, while extolling the virtues of studying sociology, teachers felt that parts of the curriculum were dated. As a result, we described the curriculum as both powerful (in its ability to enable students to see the world differently) and disempowering (when it continued to include discredited theories). In the sociology of education, we learn about the impact of the **hidden curriculum**, but this research showed how the formal curriculum also reinforces inequality.

How rigorous are social surveys?

Social researchers often feel confident that surveys are reliable and valid for the following reasons:

- If the questions are carefully worded and tested, reliability and validity is enhanced.
- If the sample is representative, reliability is enhanced.
- If the concepts are operationalised carefully, validity is enhanced.

Which sociologists are likely to use social surveys?

Generally, surveys appeal to sociologists working in the positivist tradition. This is because they usually yield objective data by ensuring the researcher stays detached and the research is value-free. Moreover, by examining the relationship between dependent and independent variables, survey data enables social researchers to test hypotheses. However, surveys are not exclusive to positivist sociologists. Realist researchers might also be interested in testing the relationship between variables to reveal how hidden structures are operating.

> **Activity**
>
> Identify the dependent variables in the following research questions:
>
> 1. *Is educational attainment influenced by the social status of family origin?*
>
> Variables: social status / educational attainment / family origin / education
>
> 2. *Is the use of food banks influenced by social class?*
>
> Variables: social class / food bank / food-bank use / poverty

 CLASSIC RESEARCH: RACIAL DISCRIMINATION IN AMERICA

Du Bois, W.E.B. (1899/1996). *The Philadelphia Negro*. Pennsylvania: University of Pennsylvania Press.

Study overview

Originally published in 1899, Du Bois' study of the living conditions of a Black community in the United States was a ground-breaking piece of sociological work that used social surveys to collect data.

Method and sample

Du Bois moved into the Seventh Ward in Philadelphia and conducted a door-to-door survey with all households in the ward. This generated responses from 5,000 people. Using this method, he reached everyone in the neighbourhood, but was limited to studying this small area of the USA. Du Bois was helped by his research assistant, Isabel Eaten, but you can imagine how long it would have taken to collect this data.

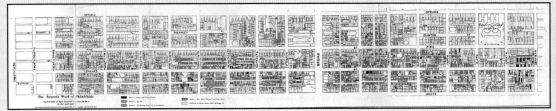

Du Bois' meticulous mapping of the distribution of African-American inhabitants of the Seventh Ward.

> **Question**
>
> Do you think Du Bois' research was generalisable?
>
> Why did he visit everyone in the ward? Could he have used a sampling method instead? Why, or why not?

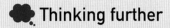 Thinking further

Du Bois based the methods of his study (including sampling) on the work of Charles Booth, who had previously mapped poverty by every street in London. It is important to reflect on the historical development of research methods here. Over time, researchers have learned that instead of having to sample every single household in each area, sampling methods can be used to ensure that research is representative but more practically achievable. For example, Booth and Du Bois could have selected addresses such as going to every tenth household in a street rather than sampling the entire location. More than this, both Booth and Du Bois could only sample specific geographical areas, whereas sampling methods enable researchers to undertake much wider – sometimes national – surveys.

Findings

The small geographical area that Du Bois surveyed included residents ranging from affluent White people and Black elites to poorer Black communities. Du Bois' findings showed that there was great diversity within the Black community, and he argued that it was wrong to characterise this population simply as 'precarious' – that is, a community in which poverty, crime and illiteracy are common. Rather, Du Bois showed evidence of significant divisions within the Black community, including the 'talented and well to do', 'labourers' and 'working poor'. Indeed, the poorest and most precarious members made up only 10% of the community. Moreover, he showed that most of the problems experienced by poorer Black communities people were a result of racial discrimination. Specifically, members of this community found themselves excluded from certain jobs and had to pay high rents for substandard housing.

While the White community held an image of the Black community as inferior, Du Bois' social survey provided evidence that this perception was invalid and discriminatory.

Evaluation

Du Bois' study was an incredibly important piece of sociological research, reliably and validly providing information on a discrete population and raising awareness of racial discrimination. Du Bois made no secret of his political interest in studying racism and was openly reflexive on his positionality as a Black academic. In fact, he drew attention to the importance of such reflexivity – he was a civil-rights activist alongside his academic work.

Although his survey was significant, with hindsight it is clear that he did not need such a large sample when gathering data.

Activity

Try extending the evaluation of Du Bois' study by applying the STRIVE criteria.

CLASSIC RESEARCH: POVERTY IN THE UNITED KINGDOM

Townsend, P. (1979). *Poverty in the United Kingdom*. London: Allen Lane and Penguin Books.

Study overview

In this classic study published in 1979, Townsend aimed to devise what he called an 'objective measure of **relative poverty**'. He wanted to show that certain levels of income and participation in society were deemed to be normal. Anyone who fell below normal levels was experiencing deprivation, he argued, even though they may not be considered poor if measured by minimum standards such as being unable to feed, house and clothe themselves.

Method and sample

Townsend made a distinction between actual and socially perceived need, between normative and individual notions of poverty. In other words, he distinguished between a conventional notion of **absolute poverty** and a more dynamic and relative account of poverty. To measure whether someone lived in relative poverty, he constructed a deprivation index (the concept) and then devised several questions to measure it (indicators). Indicators included:

- Did not have a party on last birthday (children only).
- Has not been out in the last four weeks to a friend or relative for a meal or snack (adults only).
- Does not have fresh meat (including meals out) as many as four days a week.
- Household does not have a refrigerator.
- Has not had a week's holiday away from home in the last 12 months.

Townsend's study took over ten years to conduct. His **pilot** survey in 1967 comprised a 120-page long questionnaire distributed to 150 households scattered across London. The main survey, conducted between 1968 and 1969, was 39 pages long and was carried out by trained interviewers. In total, the survey included 3260 households and involved 10048 individuals across 51 constituencies in the United Kingdom. The magnitude of the survey produced a wealth of statistics that took a long time to analyse.

> **Question**
>
> What do you think Townsend meant by an *'objective measure* of relative poverty'?

> **Questions**
>
> Do you think these are good indicators of relative poverty? Why, or why not?
>
> Are any of the measures Townsend used now redundant?
>
> What indicators would you use in the twenty-first century to measure relative poverty?

Townsend used a random sampling method to ensure his research was generalisable to the entire UK population. He used a stratified multi-stage technique because he could not sample the entire UK. At the time of his research, there were 630 constituencies in the UK, and he chose to focus his study in 51 of them. He then stratified these constituencies into high-, middle- and low-income areas. Next, he acquired a full list of addresses from the chosen constituencies and stratified them according to age and family size, so that he did not have to travel all over the country. To ensure his sample was representative of the wider population, he compared it to the official statistics. He was confident of its representativeness.

Townsend felt it was particularly important to follow up on non-respondents, believing they were more likely to be poor, such as elderly people and those with lots of children.

The response rate was high: 82% of people that the research team approached agreed to complete the survey. The 18% that did not complete the survey were proportionally more likely to be older people or couples without children.

Questions

Why do you think it took Townsend until 1979 to publish his research?

What issues might there be with using a 39-page-long survey?

Findings

Townsend's study suggested that 25% of households were living in a state of what he called relative poverty. This was significant at the time because official government data suggested that only 7% of households were living in poverty.

His results also revealed that certain groups of people were at greater risk of poverty. They included one-person households, single-parent families, large families, unemployed people or lower classes working in poor conditions, people with disabilities and elderly people.

Evaluation

Clearly Townsend's study was sociological and rigorous. It examined a core sociological concern – that of poverty and inequality – and was systematic. It yielded a representative sample, meaning that his results were valid, reliable and generalisable. Arguably, therefore, it meets the 'gold standard' of social scientific research.

When operationalising deprivation, Townsend used indicators that were useful for that time but which might now be considered out-dated. Ethical concerns were also minimised.

In applying the STRIVE criteria, you could argue that Townsend gets high marks! However, because his research was so lengthy and time-consuming, by the time it was published ten years later, the data gathered was dated and had less opportunity to make a policy impact.

 CLASSIC RESEARCH: 'TASTE' AND CULTURAL CAPITAL

Bourdieu, P. (1979). *Distinction: A Social Critique of the Judgement of Taste*. London: Routledge.

Study overview

Bourdieu was very interested in how social classes are formed and reproduced. He drew attention to the way that various forms of **capital** (**economic**, **social**, **physical**, cultural and **symbolic**) create what he called a **habitus** – an internalised structure of perceptions, habits and schemes of action that shape people's life chances and distinguish different groups. In this study, perhaps his most famous piece of work, he wanted to explore how 'taste' (things we like, our cultural capital) is shaped by our habitus and, in turn, reinforces power and social position. In other words, he wanted to look at the ways that cultural taste works to privilege some groups of people. Bourdieu identified differences between preferences for **highbrow culture**, **middle-brow culture** and **low-brow culture** that map against, and serve to distinguish, social groups.

Method and sample

Following a period of observations and interviews, Bourdieu developed a survey that was administered to 692 people living in Paris, Lille and another small town in France in 1963. This was followed in 1967 by a survey of a further 525 people, giving a total sample of 1217.

Bourdieu argued that the time lag between the two surveys had little impact on the range of responses, other than that there were some changes to music taste during this time. Both samples were similar in terms of socio-demographics, covering similar occupations, for instance, and city and provincial dwellers. However, both samples underrepresented the most deprived groups, and some groups were deliberately excluded.

> [...] the farmers and farm workers were excluded from the analysis, after a preliminary survey which showed that the questionnaire was completely inappropriate and that [...] other methods were required to identify the dispositions of a population totally excluded from legitimate culture and even, to a large extent, from 'middle-brow' culture. (p. 505)

The survey included 25 closed questions on tastes relating to interior decoration, clothing, singers, cooking, reading, cinema, painting, music, photography, radio, pastimes, etc. The results were compared with the occupation and educational background of each respondent.

For example, in the survey, Bourdieu asked his respondents to choose which of the following statements were closest to their own views:

- Paintings don't interest me.
- Paintings are nice but difficult; I don't know enough to talk about them.
- I love the Impressionists.
- Abstract painting interests me as much as the classical schools.
- Modern painting is just slapped on anyhow; a child could do it.
- I don't need to know who painted it or how.

He analysed the responses by occupation (bivariate analysis). Table 4.2 summarises his findings by percentage.

> **Questions**
>
> Are these good questions or poorly worded questions? Why do you think this? You might want to consider, for example, whether the questions allow for the full range of possibilities a respondent might want to express.
>
> What conclusions can you draw from this data?

Classes	Doesn't interest me	Nice but difficult	Love the Impressionists	Like abstract art	Modern art's not just slapped on	Do like to know the artist
Working classes	26	62	7	4	32	7
Craftsmen, small shopkeepers	17	73	5	5	44	2
Clerical, junior executives	17	65	12	7	35	8
Technicians, primary teachers	3	50	26	22	53	14
New petite bourgeoisie	4	30	32	34	64	13
Middle classes	14	56	16	14	45	9
Industrial and commercial employers	4	51	27	17	42	6
Executives, engineers	8	27	39	26	55	11
Professions	0	31	40	29	58	13
Secondary and higher-education teachers, artistic producers	4	14	39	43	75	21
Upper classes	5	31	37	27	55	12

Table 4.2: Tastes and cultural practices of classes and class fractions (%).

Findings

Bourdieu argued that 'taste is an acquired disposition to "differentiate" and "appreciate"' (p. 466). His survey established a very close statistical relationship between cultural practices, educational capital (qualifications) and social origin (measured by father's occupation). In other words, he argued that higher social classes preferred highbrow culture and succeeded in education. He also argued that those people in higher social positions determine what stands as good taste in society, and that this enables mechanisms of social distinction between groups to form.

The research showed that lower social classes are often excluded from highbrow culture, not simply by cost, but also because they feel that things like listening to classical music are not 'for them'.

Overall, Bourdieu showed that cultural choices create and reinforce social class. His study revealed the existence of **cultural hegemony**, whereby 'dominant' and 'legitimate' taste is taught and passed down to children of higher social classes through **socialisation** and serves to bestow cultural capital, power and privilege.

Questions

What theoretical position(s) do you think Bourdieu is influenced by? What makes you think this?

Evaluation

Bourdieu's survey research dramatically changed the way that social class was conceptualised and researched by sociologists. Specifically, it forced sociologists to think about operationalising social class differently, not merely in terms of economic measures, such as income or occupation, but also other forms of capital. You can see how this has been applied to sociological research in the example of the Great British Class Survey below.

This widening of social class as a concept has, however, been criticised by some sociologists. They argue that focusing on different forms of capital, instead of the economic divisions between people, detracts from the most important problem – economic inequality between classes.

Rather than blaming working-class people for having a 'poor' culture, Bourdieu's research reveals how, instead, middle-class culture dominates and shapes what stands as 'legitimate' culture.

Bourdieu's survey aimed to be representative because it selected three locations with differing population sizes. Nevertheless, there were some limitations – it was relatively small, there were some issues around representativeness, as he missed out some people, and he combined results from two separate surveys.

CONTEMPORARY RESEARCH: INEQUALITY: THE GREAT BRITISH CLASS SURVEY (GBCS)

Savage, M. (2015). *Social Class in the 21st Century*. London: Penguin.

Study overview

The impact of social class on an individual's life chances, and changes to class composition, have always been a central topic for sociologists.

Ever since Goldthorpe and Hope's (1974) ground-breaking work, which used occupation as a proxy measure for social class, sociologists have tended to use this indicator. Essentially, all occupations are classified into one of seven categories.

Social class I	Service class (higher grade)
Social class II	Service class (lower grade)
Social class III	Routine non-manual employees
Social class IV	Small proprietors
Social class V	Lower-grade technicians and supervisors
Social class VI	Skilled manual workers
Social class VII	Semi-skilled and unskilled manual workers

Table 4.3: The seven classifications of social class by occupation.

This classification is used by the government in official statistics, where it is referred to as the UK National Statistics Socio-Economic Classification (NS-SEC). This is a good example of the enduring impact that sociological research has had on the shape and collection of government official statistics.

This measurement tool has proven exceptionally useful for gathering data on inequalities in outcome (education and health, for example) by occupation. However, as Bourdieu's survey research revealed, occupation alone cannot fully capture the nuanced experience and impact of social class. Someone in a high-status occupation may not necessarily command high wages but may engage in highbrow cultural leisure activities.

In 2011, Mike Savage and colleagues, in collaboration with and funded by the British Broadcasting Company (BBC), designed the Great British Class Survey (GBCS) to capture this more complex conceptualisation of social class.

> **Question**
>
> We often think of surveys as giving researchers more freedom, but most research is externally funded, and the research team must be aware of the financial constraints and the interests of the organisations that are funding them. Savage had a great deal of latitude to design the survey, but he had to limit the number of questions to keep the cost of the survey down. What ethical issues and compromises to the reliability and validity of the study might this lead to?

Method and sample

The survey comprised 50 questions (including questions about gender, family size, ethnicity and educational qualifications). It was designed to take only around 20 minutes to complete, to reduce the commitment needed from respondents and hopefully encourage more responses. In fact, 89% of respondents completed the whole survey.

The researchers operationalised economic, social and cultural capital by developing specific indicators.

- Economic capital was measured through questions about income, housing and wealth (the value of any savings the respondents may have).
- Social capital was measured by asking questions about social networks and social ties. Specifically, the survey listed 37 occupations and asked if the respondents socialised with people in these jobs.
- Cultural capital was measured by asking the respondents about their leisure interests, musical taste, food preferences and their use of media, including television and magazines. For example, respondents were asked how often they went out to eat, whether they listened to opera, visited art galleries or went to bingo halls. They were asked which music genre they preferred to listen to (rock, world, heavy metal, country and western, reggae, folk or hip hop, for instance). In addition, questions were asked about the types of holidays the respondents took. The researchers split the choices into what they called 'highbrow' and 'emergent' (more popular and low-brow) activities.

The survey was launched on the BBC website on 26 January 2011 and was widely promoted. Mike Savage appeared on the *One Show* the night before the launch, and was watched by 4.78 million people. The survey garnered much interest, with a self-selecting sample of 161,400 people responding by July of the same year.

The research team for the GBCS was fortunate that the survey gained such a high profile, which is highly atypical of sociological research.

There were some major methodological issues with the study's sampling method. Significantly, there was a bias in the type of people that responded to the GBCS: the majority came from well-educated social groups and those in professional occupations – arguably the BBC's main audience. To overcome this problem, the researchers supplemented the survey with another study (referred to as the GfK sample, after the research firm that carried it out). This used the same questions but sought responses from another 1026 people from underrepresented groups. The researchers used quota sampling to select their respondents.

Findings

The results of the study enabled the researchers to identify seven major social class groupings. They argued that this provided a more detailed way of understanding social class than occupation alone and was more sophisticated than a simple middle- and working-class binary.

> **Question**
>
> Bradley (2014) criticised the survey design, suggesting that the markers of cultural capital were highly selective and led to a negative view of working-class culture. Do you agree? Why, or why not?

Savage's seven categories, from left to right: precariat, traditional working class, emergent service workers, technical middle class, new affluent workers, established middle class, elite.

	% GfK	% GBCS	Description
Elite	6	22	Very high economic capital (particularly savings), high social capital, very high highbrow cultural capital
Established middle class	25	43	High economic capital, high status of mean contacts, high highbrow and emerging cultural capital
Technical middle class	6	10	High economic capital, very high mean social contacts, but relatively few contacts reported, moderate cultural capital
New affluent workers	15	6	Moderately good economic capital, moderately poor mean score of social contacts, though high range, moderate highbrow but good emerging cultural capital
Traditional working class	14	2	Moderately poor economic capital, though with reasonable house price, few social contacts, low highbrow and emerging cultural capital
Emergent service workers	19	17	Moderately poor economic capital, though with reasonable household income, moderate social contacts, high emerging (but low highbrow) cultural capital
Precariat	15	<1	Poor economic capital, and the lowest scores on every other criterion

Question

Look at Table 4.4. What do you notice about the difference in class types across the two surveys?

Table 4.4: Summary of social classes. Source: Savage, M. (2015). *Social Class in the 21st Century*.

The survey data also allowed the researchers to see what types of people resided in each class category.

	Elite	Established middle class	Technical middle class	New affluent workers	Traditional working class	Precariat
Mean age	57	46	52	44	66	50
% female	50	54	59	43	62	57
% ethnic minority	4	13	9	11	9	13
% graduate	56	43	26	11	11	3
% with jobs in professions or management	63	51	35	22	31	9
% from professional or senior management families	52	41	40	19	17	5

Table 4.5: Socio-demographic correlates of seven classes. Source: Savage, M. (2015). *Social Class in the 21st Century*.

Question

What interesting data can you draw from Table 4.5? (For example, you might expect to find 50% women in all occupations, but we can see that this is not the case.)

Research tip

The BBC continues to offer a shortened version of the survey for anyone to take, and over 9 million people had engaged with this by 2015 when the official monitoring of the survey concluded. Anyone can take the survey and the class calculator immediately generates an approximate position in the seven-category class hierarchy. You can take the test here: https://www.bbc.co.uk/news/special/2013/newsspec_5093/index.stm

Questions

Social survey researchers aim for representativeness in order to generalise from their findings.

Is Savage's research representative and generalisable?

What could he have done differently to increase the representativeness of his research?

How does this approach to social survey research differ to Townsend's?

What are the advantages and disadvantages of Savage's approach versus Townsend's?

Whose research do you think is more valid and reliable?

What are the advantages and disadvantages of having the research funded by the BBC?

Evaluation

Consider how you can apply STRIVE to the GBCS:

- **Sociological:** The research took a central sociological question – how social class should be measured and understood – and provided a new and more nuanced model based on Bourdieu's work.
- **Theoretical:** The researchers took a positivist approach.
- **Representative:** Savage recognised that the GBCS was not representative, due to using a web-based, self-selecting sample. One might argue that his use of an additional survey corrected this problem. However, one critic, Mills (2014), was less favourable in his judgment. In particular, he pointed out that the analysis relied on the second and much smaller survey and, therefore, representativeness was compromised.
- **Impactful:** Undoubtedly the research enabled a sociological question to be widely debated by the public.
- **Valid and reliable:** This study may be regarded as a valid measure of social class because it did not rely on economic indicators alone. However, Mills (2014) also criticised the study's validity by suggesting that a survey cannot ever capture the complexity of cultural and social capital.
- **Ethical and reflexive:** Ethical questions were minimised because people had to consent to the survey and their personal data was not collected. As such, the researchers had little need to reflect on their own positionality. One ethical issue was that the research was commissioned by a private company (the BBC), which had its own agenda to collate data on social class. There is also the issue of the double hermeneutic (see Chapter 1), as this research arguably changed public views about social class.

> **Activity**
>
> Compare Bourdieu's sample and study with Savage's Great British Class Survey. Why do you think people who are more deprived do not get captured by survey methods?

CONTEMPORARY RESEARCH: EDUCATION: 'SETTING' AND SELF-CONFIDENCE

Francis, B. et al. (2020). 'The impact of tracking by attainment on pupil self-confidence over time: demonstrating the accumulative impact of self-fulfilling prophecy'. *British Journal of Sociology of Education*, 41(5), pp. 626–642.

Study overview

Francis and colleagues used survey research to examine the impact of placing students into ability sets for mathematics and English (attainment tracking) on their self-confidence. The researchers were interested in finding out whether the **self-fulfilling prophecy** in education was true – that is, whether students who are placed into lower subject sets ended up faring worse than their peers placed in higher sets.

Francis et al. believed their study to be important because social researchers have suggested that attainment tracking and setting is socially unjust. They claim that it creates an uneven playing field for students and can heighten, rather than bridge, social inequalities in education based on factors such as social class, ethnicity, disability and gender. Research already reveals that students from lower socio-economic backgrounds, as well as minority ethnic students, are more likely to be placed in lower-ability sets.

In this research, Francis et al. were particularly interested in examining the impact of setting on student self-confidence, and whether the effects of this were long-lasting.

Method and sample

To undertake this study, the researchers used a longitudinal survey. They approached 1,006 schools in the UK, and finally recruited 126 schools through first advertising the study and asking for volunteers, then following this up with cold-call sampling using a stratified random sample of schools.

The schools were distributed across England and were broadly representative. The researchers only included schools that were non-selective by attainment (not grammar schools, for instance) and state-funded (excluding private schools). Further criteria for inclusion were that the school already set students in mathematics in Years 7 and 8.

All schools in the study were given surveys for students to complete. However, the analysis focused on those students who experienced setting in two subjects: mathematics and English.

Questions

How would you describe the two sampling methods used?

What problems can you identify with the sampling methods used in this research?

The survey was distributed twice. Once at the beginning of the study and once two years later. Over the two years, 9,059 students were included in the study. Out of these, 6,167 students had been set for mathematics and 2,892 for English.

When collecting data on students' setting and self-confidence, the researchers generated demographic data, including information on gender, household socio-economic status and ethnicity. For socio-economic status (social class) they used the proxy measure of parental occupation, taking the details of the parent with the highest occupational status, classified during the analysis into three occupational categories: professional/managerial, intermediate and semi-skilled/unskilled.

> **Questions**
>
> Why do you think the researchers wanted to gather this demographic data?
>
> Are there any problems with how the researchers' operationalised social class?
>
> Do you think there are any limitations to drawing conclusions over a two-year time period? If so, what are they?
>
> Can this be described as a longitudinal study? Why do you think this?
>
> What would make these research findings more robust? Explain your answer.

Teachers administered the surveys, and students then completed them online. The questionnaires took around half an hour to complete and included:

- perceptions of mathematics and English
- liking for school
- perceptions of attainment grouping.

Self-confidence measures were drawn from surveys that had tested these measures in other research to ensure validity.

Different schools took different approaches to putting students into sets. Some schools had three sets (top, middle and bottom) and others had four sets of ability ranges. To code the data for analysis, the researchers grouped all the sets into three (top = 1; middle = 2; bottom = 3). In schools with four sets, the researchers allocated the two middle sets as 2. Researchers also accounted for movement between sets (around 20% of students) in their analysis.

Findings

The researchers found significant differences in self-confidence between students in the top and bottom sets after two years. When they compared these self-confidence levels to students in the middle set (which acted as a control group), they found that students in the top set showed a significantly higher level of confidence and those in the bottom set showed a significantly lower level of confidence. You can see this in Figure 4.1 on page 88 where the top set children are shown in dark purple. Their levels of self confidence in maths are higher than we would expect (this was measured using Hedges' *g* which is a statistical test that

> **Questions**
>
> Why do you think using harmonised self-confidence measures (scales) from other renowned international studies (rather than the researchers making up their own) increases the validity of the study?
>
> What are these sorts of questions called?

4 SOCIAL SURVEYS

measures the effect of an intervention). Children in the top set for maths also had high levels of self confidence in other subjects.

The researchers note that the gap in confidence levels was already there at the start of the study but had widened by the end of the two years.

The findings also showed that for mathematics, putting students in the bottom set not only decreased their self-confidence in that subject, it also had a considerable impact on their confidence in other subjects. Francis et al. suggest that this reveals the impact of **labelling** on students, and the creation of a self-fulfilling prophecy. When someone is labelled as a bottom set student, their self-confidence is shown to decrease, and thus their attainment does not improve, thereby meeting the expectations of being at the bottom in the first place.

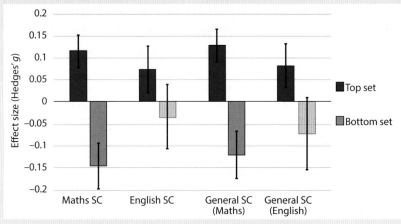

Figure 4.1: Trends in self-confidence over two years when comparing students in top set and bottom set with those in middle sets. Source: Francis, B. et al. (2020). 'The impact of tracking by attainment on pupil self-confidence over time: demonstrating the accumulative impact of self-fulfilling prophecy'.

The findings for students placed in English sets was still significant (see Figure 4.1), but not as significant as in mathematics. There was still a marked increase in self-confidence for those in the top sets over the two years, but less so than in mathematics. There was also a decline in self-confidence in those in English bottom sets, though again lower than that recorded in mathematics.

As well as controlling for social class and ethnicity (to ensure that it was subject setting rather than other factors that increased or decreased self-confidence), the researchers also ran statistics to control for prior attainment. That is, they considered whether students who were placed in lower sets for mathematics and English already had lower levels of self-confidence before they were placed in sets at secondary school. While there was some evidence of lower or higher confidence levels based on prior attainment, the researchers showed that setting itself still had an independent statistically significant impact on self-confidence levels.

> **Question**
>
> What other factors might impact a student's confidence in subjects such as mathematics and English besides those identified above? You could consider factors such as absenteeism or exclusion, for instance.

Labelling students as low or high achievers by placing them in sets not only impacts how students feel about their own capabilities for learning, it also impacts the way that other stakeholders – such as teachers, and possibly peers and parents – interact with children around their learning. Francis et al. suggest that this internalisation of labelling will likely have a strong influence on children's self-belief in their educational journey and thus their future successes in education and the wider world. Interestingly, they argue that rather than a self-fulfilling prophecy, attainment tracking constitutes a **snowball prophecy**. By this they mean that when students are labelled through setting (or other examples of attainment tracking), the effect is cumulative, increasing over time, exacerbating inequalities between students.

> **Key concepts**
> **snowball prophecy**
> a phenomenon in which the labelling of students has a cumulative impact over time, becoming more significant to a student's self-identity

Evaluation

The researchers took a systematic approach to sampling, aiming to be representative. Their sample size was large and the response rate was good. Their use of harmonised questions increased the reliability of the study. Tracking different groups over time allowed them to make comparisons. Because they tried to control for other variables, such as social class, ethnicity and prior attainment, we can be confident that the findings are valid.

Much sociological theory talks about labelling and the self-fulfilling prophecy, and this empirical study was able to show evidence of both concepts, as well as inductively generating a new concept – the snowball prophecy. Overall, this study scores well according to STRIVE rules.

👮 CONTEMPORARY RESEARCH: CRIME: 'PULLED OVER'

Epp, C. R., Maynard-Moody, S. and Hader-Markel, D.P. (2014). *Pulled Over: How Police Stops Define Race and Citizenship*. Chicago: University of Chicago Press.

Study overview

Drawing on an extensive survey of drivers, this research was designed to assess whether some people are more likely than others to be pulled over by the police in the USA, and whether the choice to stop a driver was racially prejudiced.

Method and sample

The researchers used a stratified random sample of adult drivers in the Kansas City metropolitan area (which comprises 44 cities) in 2003–04. The researchers used phone calls to contact respondents and out of an initial 8666 people contacted, they successfully achieved a sample of 2329 drivers over 18 years of age.

Using a driver survey allowed the researchers to gain a more detailed account of being pulled over. Importantly, official statistics do not provide information such as how drivers perceived the stop. The research team were interested to see if there were differences in responses given by African American and White drivers. They recognised that Kansas City was metropolitan, but they were confident that their results could be applied elsewhere, including to rural areas.

Findings

The survey revealed that 12.2% of White drivers reported being stopped, compared to 24.5% of African American drivers. White drivers were more likely to be stopped for 'traffic safety' issues. For example, 58% of stopped White drivers were given the explanation that they contravened driving rules, particularly speeding, compared to 35% of African American drivers. In contrast, African American drivers were more likely to be stopped for 'investigation': 35% of White drivers were stopped for investigations compared to 52% of African American drivers.

The researchers also found differences in drivers' experiences of being stopped, which varied depending on the reason for being pulled over. Perceptions of police conduct were described as significantly different by White and African American drivers. The two charts in Figure 4.2 below highlight these differences.

> **Questions**
>
> The researchers published this research a decade after the data was gathered. What issues might there be with this?
>
> Using the STRIVE rule of representativeness, how good is this sample?

> **Questions**
>
> What can account for the statistical gap between numbers of African American and White drivers who were stopped? Was it that African American drivers were pulled over for breaking the law more often, or might other factors account for the difference? If so, what might these factors be?

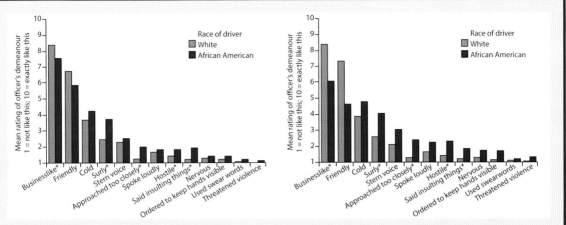

Figure 4.2: Left-hand chart: Officer demeanour, by race of driver, in speeding stops; Right-hand chart: Officer demeanour, by race of driver, in investigatory stops. *Difference by race of driver is statistically significant. Source: Epp, C. R., Maynard-Moody, S. and Hader-Markel, D.P. (2014). *Pulled Over: How Police Stops Define Race and Citizenship.*

Researchers used attitudinal questions to ascertain whether there were differences in perceptions and experiences of stops, and of wider trust in police, between the two ethnic groupings. For example, they asked:

- *How strongly would you agree or disagree with the following statements:*
  - *The local police are out to get people like me*
  - *If I needed help, I would feel comfortable calling the local police*
  - *The local police are rude to people like me*
- Respondents were also asked to agree or disagree with a range of statements:
- *How much of the time do you think you can trust the local police to do the right thing?*
- *To what extent do the police in the Kansas City area treat people fairly, regardless of race?*

Responses were recorded on a Likert scale to assess whether the drivers felt distrust or trust.

The answers to these questions showed that none of the White drivers feared being pulled over by the police – they felt able to drive in any area and believed they would be treated fairly by the police. In contrast, African American drivers were worried they would be stopped, handcuffed and arrested even if innocent.

Questions

What data can you extract from the bar charts in Figure 4.2?

What does this research infer about racial discrimination in police stops?

Question

What, if any, problems can you identify with the phrasing of these attitudinal questions?

Evaluation

The study attempted to use the gold standard random sampling methods, and it achieved a reasonably large sample/high response rate. Of course, it is hard to generalise from a study conducted in one area. There may be issues of validity due to the leading nature of some of the questions. Nevertheless, the study revealed important differences in the experience of being pulled over and evidence of racism within the police force.

CONTEMPORARY RESEARCH: FAMILY: 'MALE INFERTILITY' AND GENDER NORMS

Hanna, E. and Gough, B. (2020). 'The social construction of male infertility: a qualitative questionnaire study of men with a male factor infertility diagnosis'. *Sociology of Health & Illness*, 42(3), pp. 465–480.

Study overview

Hanna and Gough wanted to understand men's experiences of infertility – a topic area that is relatively unspoken of due to societal expectations around men's virility and normative expectations of fatherhood.

Method and sample

Their survey research was different to the surveys described in other studies in this chapter. Hanna and Gough purposely used open-ended questions to elicit qualitative data to answer questions about men's experiences of infertility, rather than collecting primarily quantitative data through closed questions.

Because of the sensitivity of the subject area and the stigma attached to male infertility, the researchers thought that men would open up more about their experiences through an anonymous survey than they would through an interview.

The researchers used 15 open-ended questions that enquired about the lived experiences of infertility. For example:

- *What would you describe as the most stressful or difficult aspect of your experience with infertility?*
- *How do you feel you have coped with infertility? What strategies have you used to help manage this situation?*
- *Please describe any positive or negative experiences with health professionals you have had during the course of diagnosis or treatment for fertility problems.*

The questionnaire was administered online via social media (Twitter) and promoted and shared through a partner charity, Fertility Network UK. The survey was anonymous, but men could provide further information if they wanted to be included in future studies. The researchers noted that 70% of men willingly gave their personal details.

The researchers noted the following limitations of qualitative questionnaires:

- an inability to follow up with participants and probe further or seek clarification on responses
- self-selecting sample
- those responding had to be able to access the internet and had to be literate.

Activity

Reflect on the researcher's rationale for using open-ended questions in survey research. What limitations might there be with this method? Consider factors such as time (to analyse the data), inability to follow up and probe respondents, etc.

Question

Do you think any of these questions could be regarded as leading or double-barrelled? If so, which ones and why?

The sample comprised 41 responses to the questionnaire and included 24 000 words of data to be analysed. Respondents were primarily from the UK, 78% of whom identified as White British. All had working- and middle-class jobs.

Findings

The data was analysed thematically, with the following themes identified.

Failing at masculinity

The researchers argued that many of the men felt that infertility clashed with the normative expectations of **hegemonic masculinity**.

Read the following extracts from the qualitative survey data:

> *I still see myself to this day, since I was diagnosed as infertile to be less than other men who can father their own children. (p. 27)*

> *Male factor infertility I think has changed my identity because I slowly became resigned to the fact that I probably wasn't going to be a parent. Being 'that bloke without children' became part of my identity. (p. 6)*

> *It's bread [sic] in our DNA to make babies. That's the purpose of sex when you are older, is to make babies. It makes me feel worthless that I couldn't have kids. (p. 23)*

> *[…] at the time the love of my life hated me because I couldn't get her pregnant as that's all she wanted from me and I couldn't give her a child. It was hard to wake up every day knowing that, I felt like I was half a man because I couldn't have kids. (p. 23)*

> *This has impacted on feeling like I have failed my wife […] and the image of how I believe people see us as a couple and I am no longer able to feel like a man. (p. 19)*

Invisibility and stigma

Invisibility and stigma were other themes that emerged from the data.

> *Fertility treatment needs to become less female-centric and concentrate on both male and female. This would help remove any stigma and encourage more men to be open and honest about their fertility. (p. 16)*

> *There are always women in the media admitting that they have had infertility issues but we so rarely hear about men. I know it is a big issue for men and the risk of being called 'less of a man' but I feel infertility is something that many men suffer in silence, fearful of being called hurtful names. It is a silent problem that many men suffer. (p. 25)*

> *Finally opening up about it has helped me deal with the situation better, realising I am not the only one. (p. 5)*

Trauma of infertility

The final theme identified was the trauma of infertility. The researchers pulled out the following quotations to illustrate this:

Question

What are some of the limitations of relying on a self-selected sample? You might recall from Savage's study that self-selection methods of sampling often lead to a biased sample.

Question

What do these statements from participants reveal about the normative expectations of fatherhood?

Question

What do you think these quotations from the qualitative survey data reveal about the researchers' themes of 'invisibility' and 'stigma'?

At first I was an emotional wreck, I struggled with depression relating to my diagnosis for the best part of 10 years. (p. 33)

It is something that I cannot get over and I am devastated. (p. 17)

The most painful aspect was realising I could never have a genetic child of my own, it was invisible grief and so difficult for anyone to understand unless they have been through it themselves [...] the idea of never meeting this child broke my heart. (p. 16)

The authors, findings suggest that male infertility is highly emasculating, distressing and continues to be stigmatised. The men in the study revealed themselves to feel like 'failures'. The dominant social construction of hegemonic masculinity (and 'reproductive masculinity'), which frames men as virile and is upheld in social ideals of fatherhood, reveals how men who cannot meet these social ideals regard themselves as 'deficient'.

Evaluation

The research was important because it gave insight into an under-researched and taboo topic area. Even though the researchers felt they could only approach men to discuss this topic through questionnaires, it is interesting that such a large number agreed to a follow-up interview. This small and self-selective sample raises concerns about the representativeness of the study. The rich level of data enhanced the validity of the findings. Even though this is a sensitive topic, ethical issues were managed because respondents chose to take part with informed consent.

Social surveys as a social research method

1. How have the methods of undertaking social surveys changed over time?
2. Why do you think fewer sociologists use social survey research methods nowadays?
3. What are the advantages and disadvantages of using open and closed questions in social surveys? Draw on the studies in this chapter to explain the insights and challenges with the different question types.
4. Look at the information in the studies on the Great British Class Survey and 'Pulled over'. To what extent is it possible to generalise from the findings of each of these studies?
5. In the surveys outlined in this chapter, which questions do you think were the most reliable and why?
6. How successful do you think the researchers were in this chapter at operationalising their concepts?

Advantages of social surveys:

- They can be quick and cheap to conduct, depending on the research design, when compared to methods such as interviews and ethnographic research. Savage's research, for instance, was cost-effective because it used the internet to achieve responses, while W.E.B Du Bois' research was likely very time-consuming because he surveyed every door in a large US district.
- They mainly use closed questions, which are easy to analyse using computer software.
- If survey samples are well designed, they are likely to be representative and therefore generalisable to a wider population. We can make judgments about representativeness because researchers usually give details about how the sample was selected.
- Surveys can produce statistical data that can be used to look at patterns, including longitudinal trends as well as relationships between independent and dependent variables.
- Social surveys are usually very reliable. They can be repeated, and anyone can administer them.
- Researchers can devise their own questions and are in control of operationalising their concepts.
- Ethical concerns are minimised.

Disadvantages of social surveys:

- Some very large studies (such as Townsend's) may be very time-consuming at the analysis stage, so the data can easily become dated.
- Survey questions can never fully capture the complexity of social life. There are three problems with this:
    - Respondents may have to choose the answer that is nearest to their experience but does not exactly fit.
    - Closed questions cannot capture individual meaning and nor can they ensure that each respondent has understood the question in the same way.
    - When operationalising concepts into indicators, we cannot always be confident that we are really measuring what we are seeking to research. For example, Savage's study was criticised because the questions did not fully capture social and cultural capital. This also raises questions concerning the validity of the research.

CHAPTER 5: INTERVIEWS

Imagine that you want to find out about why some men are choosing to become vegan, or what housework means to new mothers. You could design a questionnaire to ask specific questions, but would this allow you to generate rich insight, and would you be able to understand the wider context that shaped the respondent's experiences, or tease out the nuanced differences that characterise most life choices and chances? To achieve detailed insights into the workings of social life and the complexity of social interaction, social researchers often use qualitative methods such as interviews.

What is an interview?

An interview is a planned conversation designed to elicit information about someone's views, beliefs, experiences, attitudes, aspirations, anxieties and choices. These conversations can take place between two people – the interviewer and the interviewee – or they can be conducted in a group setting (focus group interviews).

Considering that human beings are in constant conversation with each other, how can a research interview be any different from an everyday discussion or debate? How does a research interview differ from a conversation that you might have with a friend or relative, a dinner-table dialogue, a meeting at work, or even an exchange between strangers on the train or in the street? The critical differences are that a research interview focuses on a carefully designed set of questions, takes place with a selected sample of people, and is carried out by a trained sociological interviewer.

The interviewer must aim to be flexible and open enough to put themselves in someone else's shoes, while remaining at a distance – what Weber (see Chapter 1) referred to as *Verstehen*. Therefore, when using an interview method, the researcher is very much part of the research process. While the interviewer aims to be objective, they can never be entirely detached and may influence data collection.

Much like the methods used in quantitative research, interview research is guided by a research question(s). However, rather than having a hypothesis that the researcher wishes to test, which narrows the scope of the data being gathered, the researcher using interview methods is usually more open to discovering new insights. The interviewing process can lead to new research questions and theoretical concepts. This process – where theory is generated from data – is referred to as induction. New findings are often expected to emerge, and established ideas can be both reaffirmed and challenged. For this reason, qualitative interviews are often used in the **pilot** stages of developing a research survey.

 ## Thinking further

Induction should not be confused with **deduction**, where theory is tested by the data. You will also see the term 'grounded theory' to describe the process whereby theory is derived from data, rather than theory determining which data is collected. In practice, the relationship between theory and data is usually closely intertwined (recursive), as a researcher rarely begins a study without some ideas of what they will find – it is hard to completely view the world separate from *a priori* assumptions. The point is that in qualitative methodology, the theoretical ideas and concepts in interview research are expected to be confirmed, developed and refined and/or unsettled and changed.

Key concepts
pilot an initial small-scale study to check the viability of a larger study
deduction a process that begins with a theory or hypothesis, which is then tested through the collection of empirical data
a priori knowledge that comes from theory and reasoning rather than experience or observation

 ## Interviews about housework

Ann Oakley famously studied housework in 1974, through unstructured interviews (with some guiding questions) with 40 women. These showed that most women found work in the home to be tedious and unsatisfactory (see below for details of this classic study). Likewise, Hochschild (1989) in her famous study (with Machung), *The Second Shift,* conducted interviews with 50 couples and 55 other people, including babysitters, divorcees who had previously been in two-job couples, day-care workers and schoolteachers (supplemented with observational research), during the 1970s and 1980s. From the data, she theorised a **leisure gap** between men and women and the **second shift** that women had to conduct at home after work, with significant implications for women's free time.

Garcia and Tomlinson (2020) wanted to find out whether redundancy impacted the domestic division of labour. They used a snowball sample to generate access to 20 heterosexual couples in the north of England. They then undertook 80 semi-structured interviews with this sample. In half the sample of 20 couples the women had experienced redundancy, and in the other half the men had experienced redundancy. The researchers interviewed the partners separately so that they would not influence each other's responses, and they completed the interviews six months apart to see if the domestic division of labour changed when there were shifts in work patterns, including job loss. This research could not assess the impact of class and ethnicity. However, the study suggested that women undertook the bulk of the household work, even when men were made redundant and had more time at home.

Questions
What do you think are the advantages of interview research?

How representative do you think these samples were? Give reasons for your response.

Case study: Outsourcing housework

Much research on housework looks at the fact that women do more of it, but Singha's 2019 study, *Work, labour and cleaning: The social contexts of outsourcing housework*, is interesting because it looks at the experiences of women who pay other women to do their housework in Britain and India.

Singha wanted to know about two things. First, she was interested in whether cleaners perceived their job as 'work' (using mental and/or manual skills) or 'labour' (unskilled, 'natural' women's work). Second, given the potentially exploitative nature of hiring domestic cleaners, she wanted to explore the experiences and rationales of women who called themselves 'feminists' but who outsourced cleaning to other women. Singha was interested in the hierarchies of class between these women. How did middle-class women feel about employing working-class or lower caste women to do their household chores? Did they experience conflicts with their 'feminism' and their commitment to empowering women when they hired other women to do work that is historically stereotyped as 'women's work', viewed as menial, cheap labour, done by those without educational qualifications?

Singha conducted semi-structured interviews with 27 British domestic cleaners and 24 Indian domestic cleaners and their paid employers (all feminist academics – 21 from British universities and 29 from Indian universities). She found that the interpretation of domestic cleaning as either work or labour (or possibly both) depended on the ways that cleaners were treated, as well as the wider social context in which the cleaning took place. Many of the British cleaners, for instance, regarded their cleaning as 'work' if clients were respectful (interpreted as doing some general tidying before they arrived, and not leaving toilets in a filthy state).

Singha looked at different cultural contexts (Britain and India) because in India there is a history of seeing domestic work as servitude, whereas in the UK domestic work is typically viewed as class-based but also racialised labour. She found that female academics did not consider it anti-feminist to employ cleaners, but they did try to ensure that cleaners were working, not labouring. They employed various strategies to differentiate between the two: they tended to hire independent cleaners rather than agency staff, who they believed may be exploited. Some tried to treat their cleaning staff as 'colleagues'. Some would pay more than national minimum wage or would pay their cleaners for bank holidays even if the cleaner did not work on that day.

Interestingly, Oakley was interviewed by journalist Ian Sinclair in the *Morning Star Online* in 2014 to mark the fortieth anniversary of the publication of her book about housework. She noted that gender imbalances between men and women persisted but that there was an increased likelihood that middle-class women would buy help – estimated to be about one in seven families in the UK.

> "That's not a solution because very often the people who are hired are women and they are underpaid, their job conditions and security are not good," she says. "And usually, it is the woman in the household who is responsible for looking after the hired help. So, you've simply passed the oppression on in some sense."

How do sociologists use interviews?

There are four main types of interviews.

Structured interviews

In structured interviews, the researcher follows a strict and standardised set of questions, always in the same order. In some ways, this is akin to asking a set of open questions in a survey, and they are designed carefully and ahead of the interview.

Semi-structured interviews

Semi-structured interviews also have a set of questions that the interviewer must ask, but there is an opportunity to probe (ask more questions) when an interviewee has given their first response. An interviewer might ask, 'Can you elaborate a little more?' or 'Can I ask you what you mean?' The interviewer decides in the moment if more detail would be useful and can follow new lines of enquiry if they emerge during the research conversation. The huge advantage here is that the interviewer can check that the respondent has fully understood the question.

Unstructured interviews

In an unstructured interview, the interviewer steers the interviewee in a more reciprocal and conversational way. While the interviewer will have a list of themes (prompts) to guide them, topics are discussed as they emerge, rather than being constrained by a pre-decided list of questions in a particular order. The flexibility of this approach is a key advantage, permitting new themes and questions to emerge. However, unstructured interviews can take a long time to conduct, transcribe (type up) and analyse. Because this format is more time-consuming, the researcher usually must rely on a smaller sample and therefore cannot be as confident that they have represented the full spectrum of views and experiences. They also must ensure that the interviewee stays focused and does not meander off topic.

Focus groups

Focus groups are group interviews usually comprising at least four people, and they can take a structured or unstructured format. The focus here is on eliciting the group dynamic – the discussion usually takes place around a theme – to establish the extent to which ideas are shared, whether there are differences that a group of people express on any subject, and the way that topics are negotiated and views challenged and defended. Sociologists might be interested in what type of response dominates the discussion and the ways that challenging ideas are managed.

Group discussions can foster debate, whereas individual interviews only capture a single perspective. Undertaking a focus group can also be more time- and cost-effective. In this type of interview, the interviewer takes on the role of facilitator and moderator, letting the discussion flow but also keeping the respondents on track – guiding but not interfering. Transcribing the conversation can be difficult, especially if it is hard to discern who is speaking or when more than one person is speaking at a time. Moreover, some people may find that they are silenced in the group interaction.

 Becoming a social researcher

Taking the example of housework, design a short, semi-structured interview guide and a series of focus-group prompts. Undertake each of these types of interviews with your peers and assess the usefulness and drawbacks of each method. Which method was more useful for generating insights into the topic?

Undertaking interviews

Depending on the type of interview, the sampling method will vary. The objective in interviewing is not usually to get vast numbers of respondents, but rather to ensure that the researcher has really understood the topic of enquiry. This is known as 'reaching the point of saturation' – that is, the point at which the researcher is no longer learning anything new. Qualitative research designs generally rely on purposive sampling (see Chapter 1). Interviews can be undertaken face to face or conducted over the phone or via live platforms.

Research interviews, especially those that are unstructured or semi-structured, are usually recorded and then transcribed. Transcribing involves typing up the interview verbatim – including all the pauses and 'ums' and 'ahs' – so it is a long and expensive process. It is important to record and transcribe interviews where possible, because researchers are interested in not just *what* people say but the *ways* that they articulate their ideas and views. It would be difficult to note down the content and the context of responses without recording devices. Indeed, the development of audio and video recording has been central to the expansion of this methodology.

Once interview data has been transcribed, researchers must analyse it. They typically use a form of thematic analysis that involves looking through the interview transcripts for recurring concepts and themes, which can then be coded. Sometimes researchers use computer software, such as NVivo, to assist in this analysis. It is also possible to subject interview transcripts to **discourse analysis** (see Chapter 7).

 Researcher insights

I (Sarah) undertook a research study to explore the professionalisation of complementary medicine (Cant and Sharma, 1998 and 1999). This work largely relied on data from 72 qualitative, unstructured interviews with therapists within homeopathy, chiropractic and reflexology. There were no official registers, so the sample was achieved using the snowball technique. The interviews yielded rich insights into the ways that practitioners were trying to enhance the reputation of their medical approaches. However, this sample could not capture the experiences of practitioners working within the other 160 therapies in the UK.

Although all interviewees gave their consent to be interviewed, they sometimes told me things in the interview that they did not want to be shared. Some were forceful about the consequences if I did so:

> If you dare play this to anyone I will personally come and wring your neck […] you will have me after you if you say it. (1999, p. 252)

They were also often very critical of other people I was interviewing, and were competitive, trying to impress me:

Representative of organisation: Have you interviewed X yet?

Sarah: Yes.

Representative of organisation: Oh. Where did you conduct the interview?

Sarah: Well actually we went out to lunch.

Representative of organisation: [...] well let's not stay in our office. I will treat you to tea in this nice hotel I know. (1998, p. 249)

Moreover, like Oakley, I found myself being asked for advice about how the therapists might achieve higher status and whether I could give them contacts within the university sector.

Overall, I had to think carefully about my role in the research process and had to take great care when it came to writing up the research, choosing to leave out some of the more fascinating insights. Intriguing data had to remain unused because its inclusion would have revealed the identity of the respondent.

In another study, examining social polarisation in Margate, interviews and focus groups supplemented official statistics (Koch et al. 2020). I (Sarah) undertook eight focus-group interviews that included bowling green members, youth club members, Roma mothers at the local school, a mothers' group, a local extended family and a group of local artists. Thirty-nine interviews, achieved again through snowballing techniques, involved semi-structured conversations with local council employees, social workers, street pastors, GPs, police officers, gallery representatives, artists, teachers, shopkeepers, community workers, local politicians and local townsfolk who we met in cafés, hostels and pubs and who agreed to share their experiences.

Some of the difficulties and messiness of research were revealed in this study. For instance, in one interview the respondent spoke so quietly that transcribing the interview was almost impossible. In another (not included in the 39 reported in publications), the respondent took issue with one of the questions and withdrew their consent, so the data was lost. In another, a recovering drug addict became too tired to continue after 20 minutes. A lively focus group with artists culminated in an argument between the interviewees which, while insightful of local tensions, took a lot of diplomacy to restore some semblance of order. Additionally, the youth club members were focused on showing off to one another, posturing and bragging about their experiences, and it was hard to be confident about the authenticity of their accounts.

> I was keen to make sure that I heard the voices of the local migrant community but was concerned that this would be a harder group to reach, as my own positionality and inability to speak the languages required could be a barrier. With the kind support of gatekeepers, I achieved access, and bilingual members of the community often translated for me. In this way, I gained important insights but certainly lost some of the nuance of experience through my inability to directly communicate with my respondents.
>
> These experiences reveal how interviews can be unpredictable, sometimes yielding unuseable data, and how they require diplomacy and tenacity to secure access. They also reveal the challenge of writing up data, and show that data sometimes gets left out of the final write-up. These elements are rarely acknowledged in formal articles, but they constitute the practical realities of undertaking research in the field.

> **Question**
>
> What does the Researcher insights example highlight about the tension between the *theory* of research methods and the *doing* of them?

How rigorous are interviews?

As a rule of thumb, interviews are high on validity because interview data elicits a deep and contextual understanding of respondents' accounts. However, interviews tend to be less reliable than some other methods, because they are harder to repeat and the sample size is much smaller. Of course, this depends on the type of interview you are evaluating and the way that respondents shape their answers. For example, it is difficult to repeat unstructured interview studies, but structured interviews can be undertaken by many different researchers and can be transferable to other settings and groups.

You can also judge the quality of interview data by examining the interviewing process and the role of the interviewer. Orchestrating an interview has many advantages – interviewers can build rapport with the interviewee(s) and foster a relaxed atmosphere so that the respondent can feel at ease and not judged. As such, unstructured interviews are useful when exploring difficult or sensitive subjects. That said, there is always a risk that respondents may not wish to share information, or they may provide what they see as the 'acceptable' answer. This is referred to as giving a public account rather than a private one – recounting a **socially desirable response** (similar to the Hawthorne Effect, see Chapters 1 and 8) – and can impact a study's validity.

The advantage of establishing **rapport** with an interviewee also comes with tensions. When there is scope to ask any question, there is always the risk that the interviewer might steer the conversation in a leading way. Body language, tone of voice and the positionality of the interviewer can influence how respondents answer. For example, a middle-aged, White British woman researching the experiences of school-aged British Bengali boys would be in a position far removed from her interview subjects, which would limit the insights generated.

Feminist scholars note that rapport also comes with responsibility. If you have taken the time to get to know an interviewee and have developed rapport, it is not always easy to leave them at the end of a study (**exit the field**) or refuse to offer support

> **Key concepts**
>
> **socially desirable response** in an interview, a response that the respondent thinks the researcher wants to hear and/or fits with societal expectations and norms
>
> **rapport** the trusting relationship built between a researcher and a respondent
>
> **exit the field** the process and management of leaving the research field

if a respondent seeks it. Oakley was particularly critical of sociologists who used interviews, effectively interrogating their respondents, mining them for information and then not giving anything in return. She argued that interviewers must recognise the power they wield and should ensure that the exchange is reciprocal. This could involve answering questions that interviewees ask and making their own position clear. Other methods to reduce exploitation include incentivising interviewees to take part – by offering payment, for instance. However, incentives might also mean that participation in the study and any answers given are shaped by the prospect of a reward.

Which sociologists are likely to use interviews?

Interviews tend to be associated with interpretivist, realist and constructionist perspectives. Interpretivist interviewers see the world as being made up of interactions between people and the interview is a method to uncover individual meaning; interpretivists tend to use thematic analysis. Realists may use interview data to look at individual experiences but with the aim of revealing the hidden social structures behind these experiences. Finally, constructionists would use interview data to look at the power relations articulated in and through language. Realists and constructionists are more likely to use discourse analysis methods to analyse interview data (see Chapter 7).

CLASSIC RESEARCH: ANN OAKLEY ON HOUSEWORK AND MOTHERHOOD

Oakley, A. (1974/2019). *The Sociology of Housework*. Bristol: Policy Press.

Oakley, A. (1979/2019). *From Here to Maternity: Becoming a Mother*. Bristol: Policy Press.

Study overview

Ann Oakley was keen to research topics that had been ignored by male sociologists. Indeed, she went so far as to suggest that sociology was sexist because of its male orientation. Her feminist position not only shaped what she studied, but also how she undertook her research. She was particularly critical of the fact that sociology only studied topics that men had chosen to research, while at the same time suggesting their choice of research topic was value-free and objective. Oakley argued that this made sociology inherently uncritical and un-reflexive. She therefore made the case for studying topics such as housework and motherhood, which were extremely personal to her and had been ignored by sociologists up to that point. In doing so, she foregrounded her own bias:

I am a feminist, an academic sociologist, and a woman with children. I was not a feminist until I had children, and I became a sociologist as an escape from the problems of having children. I thought it was my vocation as a woman to be a mother. When my son was 16 months old, my first daughter was born, but the time that followed was an unhappy haze of nappy washing and pill taking, as I found I could not make my dream of domestic contentment come true. I felt depressed and oppressed […] I registered to do a doctoral thesis and embarked on my research on housewives. (1979/2019, pp. xii–xiii)

This personal testimony is, of course, not intended to be definitive, there are many ways of having a baby. The point is that academic research projects bear an intimate relationship to the researcher's life, however 'scientific' a sociologist pretends to be. Personal dramas provoke ideas that generate books and research projects. (1979/2019, p. xiv)

> **Question**
>
> Oakley was open about her own political persuasion and personal interest. To what extent do you think such transparency unsettles her claims to be a social scientist?

Housework: Method and sample

In the sociology of housework, Oakley interviewed 40 London housewives, all aged between 20 and 30 with at least one child under five years old. She reflected on the fact that her sample size was 'undoubtedly small'. She wanted to focus on social class, so she selected half working-class and half middle-class women to look at the experiences within each group. To get this sample she went to two different GP practices in London, took the doctors' lists, put the names in alphabetical order, and randomly selected women. She excluded individuals based on ethnicity, choosing only to interview White Irish-born or White British-born women. She also categorised social class by the women's husbands' occupation. She then went to the houses of the randomly selected registered patients, knocked at the door and asked the women if they wished to participate.

The opening question she asked was 'Do you like housework?' She then asked the respondents how long they spent doing particular household jobs. A third group of questions related to the question of monotony and included:

- *Do you find housework monotonous on the whole?*
- *Do you ever feel that you are on your own too much in the daytime?*

> **Questions**
>
> Can you think of any ethical problems with Oakley's study? If yes, what are they?
>
> How representative do you think her sample was? Why?
>
> How effectively do you think Oakley worded her interview questions? Give reasons for your answer.

Findings

Oakley's major finding was that most women were dissatisfied with undertaking housework. Three-quarters of the sample reported that monotony was a common experience and 80% of those experiencing monotony were dissatisfied with their domestic role.

Motherhood: Method and sample

Prior to her research project on motherhood, Oakley spent six months observing expectant mothers in a London hospital. Here, she selected a sample of 66 women aged between 19 and 32 who were all due to be in the same hospital at the time of delivery. Again, Oakley excluded minoritised communities from her sample, claiming that she wanted to keep the group **homogeneous** and was aware that

> **Questions**
>
> What observations can you make about Oakley's sample?
>
> Do you think Oakley's research can be described as rigorous? Why, or why not?

reproductive attitudes varied by cultural group. She again categorised the women's social class by their husbands' occupation.

She carried out four interviews with each woman at different stages of motherhood – at approximately 26 weeks pregnant, six weeks before delivery, five weeks after birth and again at 20 weeks after the birth. She also attended six births. After the fourth round of interviews, she had 55 respondents because four women had miscarried, another's baby was prematurely delivered at a different hospital, five women had moved away, and another's marriage fell apart and she withdrew from the study.

In total, Oakley undertook 233 interviews, which equated to 545 hours and 26 minutes of recorded material.

When conducting the interviews Oakley noted:

> *You can't assume that people are merely research instruments. For, once you start to study people, it is at least a possibility that they become so influenced by the fact of being studied, that their behaviour or attitudes are changed. [...] One feature [...] is the tendency of the interviewed to ask questions back. (1979/2019, p. 299)*

In fact, her tapes revealed that respondents asked 878 questions during the course of the interviews. These included:

- Does ultrasound harm the baby?
- Can you refuse induction?
- Is it right that the baby doesn't come out of the same hole you pass water out of?
- How long should you wait for sex after the birth?

Lots of these questions were requests for information and advice. Oakley always answered the questions, noting that interviewing is a two-way process.

Oakley also asked her participants a direct question about whether her research had had an impact on their experiences of becoming a mother; 73% confirmed that it had. One of the difficulties of developing close relationships with the respondent, of course, is that it is hard to remain detached, but Oakley embraced this challenge.

> *Pauline Diggory: It's been very, I've really enjoyed it. Yes, it has helped me because I probably would have been even more worried. I mean, I think you know a lot. I mean there you are with all these different mothers, and I mean all I've got to say is, do you think Hannah's a bit sick, and you say, oh no, I've seen about so many [...] Now that just helps, just to say you've seen a few.*
>
> *Ann Oakley: But of course, I'm not a doctor.*
>
> *Pauline: Oh, I know. But I mean a doctor is not interested in a baby being sick anyway. (1979/2019, p. 303)*
>
> *Dawn O'Hara: I regard you as a friend. Somebody to speak to. (1979/2019, p. 306)*

Findings

Oakley presented her results in two ways:

- using quotations from interviewees such as the ones above
- using descriptive statistics extracted from the qualitative data responses.

Here are two examples of the statistical descriptors:

Can you describe your feelings when you first held your baby?

Not Interested	70%
Amazed, Proud	20%
Euphoric	10%

(1979/2019, p. 107)

What are the experiences of postnatal depression?

Blues in Hospital	84%
Anxiety state on coming home	71%

(1979/2019, p. 133)

> **Question**
> Why do you think Oakley included descriptive statistics as well as quotations?

Oakley's portrait of motherhood was bleak. However, she defended this because many of the women who were interviewed said that they were 'misled into thinking childbirth was a piece of cake, and that motherhood a bed of roses' (p. xvii).

Evaluation

STRIVE can be applied here to evaluate Oakley's studies. (You might also find it useful to refer back to the qualitative judgment criteria outlined in Chapter 1.)

- **Sociological:** Oakley's work was important, as she opened the **black box** of the home and motherhood to sociological scrutiny and provided a deep understanding of domestic inequality and what it is like to become a mother. Oakley extended the boundaries of existing sociological interest by studying women's experiences in a male-dominated society, as well as within the field of sociology itself.
- **Theoretical:** Her research can be described as realist because she draws on feminism to reveal the hidden and oppressive power structure of patriarchy in her studies of both housework and motherhood.
- **Representative:** There are some problems with representativeness because the samples in both studies were small. They also both excluded certain categories of women (ethnic minorities).
- **Impactful:** Oakley's research was designed to challenge existing male-dominated social structures and to bring about social change in the form of equality for women. Her research on housework raised awareness of women's experiences, but ultimately it did not change the domestic division of labour.

Oakley herself made this observation, noting that 'gender ideology continues to act as a powerful driver of behaviour' and 'there is very little evidence of change in women's and men's share of unpaid domestic and care work' (Oakley, 1974/2019, p. xi). Her work on motherhood was more impactful. She described it as the most popular book she had ever written, commenting that it generated much media attention and had been used for educating professionals about motherhood, around women's experiences of depression and anxiety. Nevertheless, Oakley's work was representative of the group she chose to study.

- **Valid and reliable:** In capturing women's own voices, Oakley was confident that the research was valid. In a follow-up motherhood study, she managed to re-interview some of the original respondents 37 years later (Oakley, 2016). These women provided 'strikingly' similar accounts of motherhood, suggesting that the original research was reliable.

- **Ethical and reflexive:** In the housework study, Oakley accessed her sample using GP lists and then went to knock on the doors of the addresses she selected. Today, researchers must abide by rules of data protection, so this type of sampling method would not be possible or viewed as ethical. All the women who were approached agreed to participate in the research. Does this raise any issues around informed consent and coercion? As you have seen, Oakley was highly reflexive about her own positionality and saw this as essential to the research process.

Case study: Interviewing first-time dads

Miller's interview research into fatherhood, 'Falling back into gender? Men's narratives and practices around first-time fatherhood' (2011), provides a contrast to Oakley's insights into first-time motherhood. Miller interviewed men longitudinally, between 2005 and 2009, about their transitions to first-time fatherhood. Like Oakley, she interviewed men at different stages of fatherhood – the first interview was 7–8 months into the pregnancy. The second interview was 6–12 weeks after birth. The third interview was 9–12 months after birth, and the fourth was around the child's second birthday.

The interviews were semi-structured, asking men questions about their expectations of birth and fatherhood, their fathering experiences, how they perceived themselves as fathers, their role as caregivers, as well as questions about their work. The men were asked to describe how they had felt when they found out they were to become a father, and subsequent interviews asked the men to describe what had happened since the last interview.

Miller's sample comprised 17 men, all White, employed, heterosexual and living in dual-earner households. Ages ranged from 24 to 39 years, with an average age of 33.7 at the time of the first interview. Her sample was achieved through advertising in different locations, including shops, leisure centres and workplaces. Interviews lasted 1–3 hours and were recorded.

The data revealed that, in the early days of fatherhood, men hoped that parental duties would be equally shared. Respondents noted that while there were differences in men's abilities to do certain aspects of parenting (especially feeding if women chose to breastfeed), they hoped to be as equitable as possible with their partners and to take on equal caring responsibilities. However, the longitudinal data revealed the impracticalities of managing and sustaining this equity once men had to return to work full time after paternity leave, which for most men was two weeks after the birth. Men found it difficult to disrupt the normative gendered behaviour of parenting because of structural factors like inflexibility at work. As Miller concludes: 'Whilst taking turns in aspects of caring, and glimpses of doing gender differently, are discernible across the data, for the most part it remains the mother who is left holding the baby.' (p. 1107).

CONTEMPORARY RESEARCH: CRIME: GYPSY/TRAVELLER CULTURES AND DOORSTEP FRAUD

Phillips, C. (2019). 'The trouble with culture: A speculative account of the role of gypsy/traveller cultures in "doorstep fraud"'. *Theoretical Criminology*, 23(3), pp. 333–354.

Study overview

In this study, Phillips critically examined the assumptions associated with 'doorstep fraud' – an offence often associated by the public and criminal justice system with the Gypsy/Traveller community, members of which are often depicted as 'rogue traders' and 'cowboy builders'. Doorstep fraud includes offences where someone offers to do work on a house or garden without having the required professional qualifications, when the job does not actually need doing, or taking the money but not doing the job. It often includes the targeting of vulnerable and elderly groups.

Phillips was interested in exploring the role of culture and ethnicity in offending rates, but more broadly she wanted to examine whether certain groups were more likely to be perpetrators of this crime and/or labelled as criminals.

Method and sample

She selected her sample by approaching the Lead of the National Doorstep Crime Project (National Trading Standards Board). She used her professional networks and emailed all 166 Trading Standards departments to generate her sample. An estimated 27 men and one woman were imprisoned for doorstep fraud and deception offences in England and Wales during the time she conducted her study.

Phillips was unable to interview some prisoners for several reasons. In some cases, the prison governors denied her access; in others she could not locate the prisoners or the prisoners themselves declined to be interviewed. She conducted interviews with five White British males and seven members of the Gypsy/Traveller community (comprising mixed White and Gypsy/Traveller origin), all imprisoned for activities relating to doorstep fraud. Her final sample, therefore, comprised 36% of the prisoners (12 respondents).

Phillips noted that her sample was small, but she argued that this was justified because a) the size of the field was small, since there were few reported cases of doorstep fraud (it is a low reporting and low prosecution offence), and b) because 'saturation' was reached after the tenth interview (no new information was being revealed through the interview data).

Phillips also recognised that other methodologies, particularly ethnographic observation, would have given an even more insightful picture into doorstep fraud, but this was impossible to conduct. As such, she felt that interviews with prisoners were the only viable way of generating insightful data into this topic area.

Questions

How representative do you think Phillips' study is? Why?

Do you think the interview method was the best choice for this study? What were the advantages and disadvantages of using this method for studying this topic area?

What other methods could she have used and what distinct types of data might this have yielded/different challenges may she have encountered?

Findings

Most prisoners interviewed associated doorstep fraud with the Gypsy/Traveller community, including the Gypsy/Traveller research participants themselves. Many of the White prisoners sampled reported that they were associated with Gypsy/Traveller communities when committing doorstep fraud. Phillips' analysis focused on interpreting the data to reveal how the cultural constructions of Gypsy/Travellers as criminal was shaped by wider discourses of racism. Her main points of analysis were as follows.

- There is a long history of criminalising Gypsy/Traveller communities.
- There is much discrimination against and exclusion of the Gypsy/Traveller community, which is negatively depicted in wider culture and particularly in the media.
- This history of **criminalisation** underpins contemporary forms of discrimination among the police and Trading Standards agencies towards these communities.
- The communities themselves often take on the labels of criminality that are given to them, which explains why the interviewees described themselves as doorstep fraudsters.
- While this label is true of a small minority of the community, it is problematic that the label is applied to the whole community, the vast majority of whose members are not criminal.
- Those interviewed noted that a nomadic lifestyle made it difficult to secure other types of work and gave them the opportunity to commit doorstep crimes, because they moved around and were therefore unknown in new communities.

> **Question**
> What theoretical perspective do you think Phillips takes? What makes you think that?

The data Phillips revealed was full of contradictions, shaped by competing power relations. On the one hand, she revealed how the data was itself a social construction that reflected the wider historical discrimination of this community. On the other hand, she took the data as fact, as a real reflection of the actual crimes, albeit underpinned by structural disadvantage. Something sociologically interesting can be extracted from this contradiction: namely, that the labelling of a particular community as criminal results in a self-fulfilling prophecy.

> **Question**
> How could Phillips' work be used to critique official statistics on doorstep crime?

Evaluation

Phillips' study is sociologically important, because it aims to reveal how certain cultures are demonised and blamed for particular types of crime. In the case of doorstep fraud, Gypsy/Traveller communities are constructed and labelled as criminal. At the same time, Phillips notes the self-fulfilling prophecy, whereby those labelled as criminal accept this identity and go on to commit crime. More than this, she identifies structural factors that explain why members of Gypsy/Traveller communities might engage in this crime, not only because of their nomadic lifestyle but also due to an inability to secure alternative income sources. In giving her paper the title 'The trouble with culture', she draws attention to the nuanced way in which culture shapes criminal activity as well as the problems with **cultural determinism**.

While Phillips' sample size was small, it was representative of that group and she did reach saturation. Her interview data is also likely to be valid – as the respondents were imprisoned for the crimes committed, they were honest in their accounts. However, there are ethical dilemmas associated with conducting research with vulnerable groups such as prisoners.

CONTEMPORARY RESEARCH: EDUCATION: DEBUNKING EDUCATIONAL STEREOTYPES OF BRITISH MUSLIM CHILDREN

Shain, F. (2011). *The New Folk Devils: Muslim Boys and Education in England*. Stoke-on-Trent: Trentham Books.

Shain, F. (2021). 'Navigating the unequal education space in post-9/11 England: British Muslim girls talk about their educational aspirations and future expectations'. *Educational Philosophy and Theory*, 53(3), pp. 270–287.

Study overview

Through interview research, Shain turned the sociological spotlight onto the experience of British Muslim children in education, not least because official statistics show that they underachieve, and this underachievement had not been subject to systematic inquiry.

In the first study (2011), Shain focused on the experiences of British Muslim boys, who she described as 'the new folk devils' in the public imagination. She argued that British Muslim boys were once typically viewed as 'passive, hard-working and law-abiding', yet after 9/11, they were recast and stereotyped as 'volatile, aggressive hotheads who are in danger of being brainwashed into terrorism, or of would-be gangsters who are creating no-go areas in English towns and cities' (abstract).

In the second study (2021), Shain interviewed British Muslim girls about their educational aspirations and future expectations. She noted that young British Muslim girls, particularly from Pakistani and Bangladeshi backgrounds, were more likely than ever to succeed in school and go on to study at university, despite being framed in the public imagination as having lower educational goals. However, this educational success does not translate into labour market advantage. Research indicates that British Muslim girls are three times more likely to look after the home and family than to take up paid work. In policy documents, these labour market outcomes are often attributed to Muslim women's cultural values, and there is a failure to acknowledge structural barriers such as work-based discrimination and expensive childcare.

Study 1: Method and sample

The research was conducted between May 2002 and October 2003. Shain used individual and focus group interviews with 24 working-class British Muslim boys aged between 12 and 18. Seventeen of them were British Pakistani, four British Bengali, two British Afghani, and one British Turkish. Shain used a convenience

Questions

How could Shain have gained a more representative sample? What might have been some barriers to achieving this?

sample (see Chapter 1) at one school and one youth group. She completed the focus group in the youth group and undertook interviews with boys in school.

Findings

Shain found the following from her interviews:

- The boys were placed in lower sets at school.
- The boys had widespread experiences of racism, ranging from name-calling to discriminatory labelling, where they were treated as 'problems' within the school. They were often called 'terrorists' or 'Bin Laden', for instance.
- Most participants self-defined as Muslim and asserted strong Muslim masculinities to counter the anti-Muslim racism they experienced in schools.
- The boys were not allowed to stand in groups of more than three at school because teachers then described them as being in a Muslim 'gang' and viewed them as threatening.

Abid: They say like ... all the people think there's a gang of Asians and they're just friends and one of the English boys who's with us, he's getting abuse off a boy and this guy in the top group he's saying to this [other] guy 'stop it' and he's saying 'oh, what you going to do? Going to get your gang of Asian boys onto me...?'

FS: Really?

Abid: The teachers even think that ... they saying, oh er, like before a few months ago we were allowed, there's this classroom there and we were allowed to sit, well not sit, but stand [outside it] but now it seems that teachers are moving us away from there.

FS: Why do you think Asian lads are being moved on?

Abid: Because they're said to cause most of the trouble.

FS: And are there Asian gangs?

Abid: No not really.

Arif: Teachers take the white people side more.

Wahid: Cos like they don't split them up. They let them be in their gangs ... they always split us up so that only one or two of us....

FS: Why do you think that happens?

Wahid: Cos they favour the white people ... it obvious, isn't it?

(p. 84)

> **Question**
>
> What themes would you extract from reading this exchange?

Shain used the interview data to generate themes. These enabled her to challenge the prominent representation of Muslim boys as a 'problem' and 'underachievers', and to instead reveal how their experiences in education were heavily shaped by systemic racism. Her research reveals how the boys used their own agency to challenge and resist these racist experiences.

Study 2: Method and sample

In her research into the experiences of British Muslim girls, Shain interviewed a total of 77 young women between September 2017 and May 2019. These included 19 British Muslim girls and women between the ages of 13 and 19 in school and further education, four women in their early 20s in higher education, and focus groups with girls aged 13–16 in secondary schools. All the participants in the research were either practising Muslims or had family connections to Islam even if they were not practising Muslims themselves. British Muslim ethnic backgrounds included Pakistani, Bangladeshi, Afghani, Turkish, Somali, Moroccan and Lebanese.

Below is an excerpt that describes Shain's fieldwork sampling:

> The main fieldwork was carried out across two geographical areas: a northern town and a southern city. In each location, focus groups were conducted in two mixed secondary schools. Interview participants were recruited via a poster call with the support of schools and colleges in both geographical locations. The poster was also shared via social media and attracted two young women who were undertaking further education in the West Midlands. (p. 276)

Questions

What sampling method does Shain use?

Why do you think she chose to use this method?

What are the advantages and disadvantages of this method?

Findings

Shain's research into British Muslim girls revealed the following findings:

- Despite research suggesting that ethnic minority girls often have lower educational goals, poor language skills and are lacking in agency, her female British Muslim sample had high aspirations for attending university.
- The girls showed high levels of agency. The interviews revealed many references to 'working hard', 'being independent', being 'driven' and 'wanting success'.
- British Muslim girls and young women were more driven towards further education than settling down with a family. This contrasted with previous research into earlier generations of British Muslim and British South Asian girls.

Evaluation

Shain focused on an important sociological question – the relationship between ethnicity and educational performance. Her research debunked many cultural myths about, and dominant representations of, the British Muslim population in the UK and, importantly, pointed to evidence of educational achievement within this community.

Her sample of British Muslim boys was significantly lower than her sample of British Muslim girls. The samples in both studies were predominantly working-class, and arguably her findings failed to capture middle-class educational experiences and trajectories. The former study was less demographically representative than the latter but both score highly on validity. Nevertheless, Shain also reflected on the influence of her positionality on the research. She gave an example of a focus group interaction in which one respondent displayed hyper-heterosexual behaviour and another tried to censor him. Shain noted that these censoring attempts may have been due to the fact that she herself was present in the group, as an older British Muslim woman and 'sister'. There are also ethical issues with studying children under 16, although Shain received informed consent for their participation.

Question

What theoretical perspective(s) do you think underpins Shain's research?

CONTEMPORARY RESEARCH: FAMILY: WHY MIDDLE-CLASS MUMS STOP PAID WORK

Orgad, S. (2019). *Heading Home: Motherhood, Work, and the Failed Promise of Equality.* New York: Columbia University Press.

Study overview

Orgad suggested that a woman's decision to end her career after becoming a mother was often framed in the media and by feminists as regressive – as a 'step back' for highly educated women. Orgad was interested in exploring how women who had decided to leave work after having children felt about this decision.

Method and sample

Orgad interviewed 35 London-based, middle-class professional women who left employment after having children while their husbands continued to work in high-powered jobs. She also interviewed five male partners who remained in paid employment.

Locating women for the study was not straightforward. As Orgad's focus was on women who had left work, she could not approach workplaces to find her sample. Therefore, she used various mechanisms of snowball sampling, such as posting notes on school mailing lists in middle-class and upper-middle-class neighbourhoods in London. She also left messages on different social media mothers' groups in London, and used notice boards in local libraries, community centres and leisure/sport centres in those areas where she thought there was likely to be a concentration of highly educated and professional women.

Orgad noted that her purposive, rather than representative, sampling method yielded a sample of mainly White women, with only one Black woman and three mixed-race women. Nevertheless, she stopped at 35 interviewees because she claimed that she had reached saturation point.

Orgad reflected on the value of interviews as opposed to other methodologies, and stressed her own commitment to the interpretivist approach in sociology:

> I see the interview as a place where people make judgments and use tools as they attempt to make sense of their place within society. I wanted to listen to the voices of women who had left paid employment, not in order to gain access to some unmediated or authentic truth but to understand how they accounted for the decision to quit their job and the consequences of this decision for their lives and identities. (p. 19)

Question

Do you think the uneven sample of men and women posed problems for Orgad's study? If so, what problems and why?

Question

In what other ways might Orgad have been able to collect useful data?

> **Questions**
>
> Drawing on the information from the study into Oakley's interviews with women and Orgad's study here, what challenges do you think such intimate interviewer-interviewee relationships pose?
>
> What are some of the advantages that such close rapport affords?

Orgad used an interview guide, which included questions related to three key themes:

- women's experiences in paid employment
- the decision to leave paid employment
- their lives since leaving paid employment.

Orgad also reflexively explained her own positionality, which was different to the women she studied. She was concerned that her own decision to stay in full-time employment while raising children may unwittingly be perceived as judgmental of the women she was studying, and that the interview encounters might be defensive or antagonistic as a result. However, this turned out to not be the case. Indeed, Orgad's interviewees often described the interview as being 'like therapy'.

In reflecting on the importance of recording the interviews, Orgad noted that 'the interviews were audiotaped and transcribed verbatim, including pauses, laughter and word repetitions, all of which proved especially important in revealing interviewee's emotions and moments where discourse seemed to fail them' (p. 21).

In the following extract, Paula – a former lawyer and a nine-year stay-at-home mother of two – is talking about the challenges she experienced while working alongside raising children.

> *I would be really often in a foul mood when [laughs] ... at the end of the day. Because I was just exhausted, you know, mentally exhausted. Yeah, it was ... it was ... I wasn't [pause] ... I'm not a natural kind of, um ... I'm not [pause] ... probably not as patient as I should be with young children ... so I did find it, yeah, quite, um [silence] wearing. It's very difficult, isn't it? Your feelings are so ... my [pause] ... it's so hard to sum it up. (p. 70)*

> **Question**
>
> What do you think this full transcript reveals?

Orgad explained that the pauses and stutters show that Paula felt inadequate, self-judgmental and unable to fulfil the socially upheld expectations of working mothers. The good work-mothering fantasy holds that mothers can switch off from work and can patiently spend time with their children.

Findings

Orgad's key findings were as follows:

- The 'return to the home' of highly educated, previously professionally employed women could not be explained as a simple result of a 'nostalgic embrace of conservative gender roles' (p. 193).
- Multiple factors shaped women's decisions to leave paid employment:
  - work cultures and work structures that did not support/enable family life
  - denial of part-time working requests
  - gender pay gaps
  - precarious work contracts
  - unsuitable childcare arrangements
  - social perceptions of mothers who work.

- The notion of the 'work-life balance' is a construct by which women measure themselves. They aim to be successful career women as well as perfect and fulfilled mothers. When they are unable to fulfil both roles, they see themselves as failures both at work and at home. Rather than regarding their failings as a result of an impossible and inconsolable balance between demands of work and family, the women interviewed internalised the contradiction, subjecting themselves to heavy critique. Orgad's research reveals how 'the seductive ideal of the balanced woman helps to mute and cover up the structural conditions that prevent women from balancing the two spheres' (p. 196). Women's decisions, then, could not simply be interpreted as individual choices.
- The long hours demanded by many professional workplaces makes it difficult, 'if not impossible' for women and their partners to participate in and balance with family life. The men interviewed revealed how they would not see their children during their waking hours in the week because of travel required for work and/or early start and late finish times. Men wanted to be more involved in parenting, but there was a necessary 'compromise' for families to manage the competing demands of work and raising a family.
- Many husbands, additionally, downplayed the work involved in mothering, instead viewing it as leisure. Some male interviewees expressed resentment towards their wives, commenting on their 'time off' during the day with questions such as 'How was your holiday today?' or 'Who have you had coffee with today?'.
- Interviews revealed that women's and men's workplaces were not proactive in discussing how women could return to work after having a baby or manage their workload flexibly.

Evaluation

Orgad's study shines a light on the social inequality that women still experience in taking on the dual burden of trying to balance domestic life with work life. Her focus on self-selecting middle-class women who could afford to 'head home' gives insight into the compromises that women make, but it does not give detail about what happens when women cannot afford to give up paid employment or when women are wealthy enough to hire help such as nannies or to enrol their children in expensive nurseries. Orgad's work stands as an insightful snapshot rather than a representative discussion of the ways in which women navigate their work and mothering.

CONTEMPORARY RESEARCH: FAMILY: DEBUNKING MYTHS ABOUT BRITISH CHINESE FAMILIES

Lau-Layton, C. (2014). *British Chinese Families: Parenting Approaches, Household Relationships and Childhood Experiences*. London: Palgrave Macmillan.

Study overview

Lau-Layton observed that the Chinese population in the UK was often depicted as a homogeneous group that followed the same traditional values and customs. These cultural depictions were drawn on to explain the educational success of British Chinese children. However, there was very little sociological research into British Chinese families and the objective of this study was to explore the diversity of Chinese family life and to capture the voices of children who are often not included in research studies.

Method and sample

Taking an interpretivist approach and drawing on constructionist theory, Lau-Layton wanted to understand parenting practices, socialisation and the types of parent–child relationships that characterised British Chinese families, as well as the freedoms (agency) that British Chinese children exercised.

As a second-generation, British-born Chinese citizen, Lau-Layton acknowledged that she held an **insider status**. This positionality gave her access to the British Chinese community and a deep understanding of the interviewees' situation and experiences. At the same time, she was aware that she had to make sure that she 'understood beyond' and 'listened beyond' her taken-for-granted understanding. To enable this, she kept a reflexive diary and cross-checked her findings with others to make sure that she was not drawing too closely on her own expectations.

To undertake her research, Lau-Layton contacted 12 British Chinese families in the north of England. Seven families were contacted via Chinese organisations and the remaining five were located through snowballing from the initial contacts. In all, 72 semi-structured interviews were undertaken by talking to one parent and one child (aged between 12 and 15, including eight girls and four boys) in each family, on three separate occasions. The 12 families had been resident in the UK for different lengths of time, some being migrant families and others British born, so Lau-Layton was confident that she had captured the community's diversity.

The researcher deliberately visited the families three times so that the interviewees could reflect on their previous answers. All participants were paid for their time and most of the interviews were undertaken in the family home to increase participant feelings of control. Nevertheless, Lau-Layton recognised that the children might have felt concerned that their answers could be overheard by their parents, so she assured all respondents that their accounts were confidential and would not be shared with other family members:

> **Question**
>
> Why do you think British Chinese family life is under-researched? You might want to reflect on the fact that many sociologists take the White British family as the norm and only focus research on minoritised communities when there is deemed to be a social problem to explain.

By outlining from the start of the project that the interview accounts were confidential, it was hoped that this would enable children to provide a true reflection of their childhoods without fear that parents would punish or be angry with them. (p. 79)

All interviews were recorded, and Lau-Layton then chose to selectively transcribe parts of the interviews (undertaking a full transcription would have been too time-consuming). Therefore, she deliberately chose sections of the interviews to analyse.

Findings

The research revealed a nuanced picture of British Chinese family life, and enabled important questioning of many stereotypes that are held about Chinese culture. Overall, Lau-Layton drew attention to the following.

- **Parenting approaches:** Some (but not all) British Chinese parents exercised considerable influence over their children's life, and placed much emphasis on patriarchal authority and filial (child) obedience. There was an expectation that children would comply with parental orders and would be disciplined (sometimes, but not always, including **corporal punishment**) if they were disobedient. However, there was evidence that parenting was more relaxed than during the previous generation, and only a minority of children described their parents as too strict. The assumption that Chinese parents were authoritarian was shown to be inaccurate, as there were far more nuances found within parenting approaches.

- **Parent-child intimacy levels:** There was evidence of generational change. Many parents worked hard to build positive, strong, warm and close relationships with their children, and this was regarded as a distinct change from their relationship with their own parents. Parents also increasingly accepted the 'westernisation' of their children and were happy for them to have diverse friendship groups and to seek romantic relationships with peers who were not British Chinese. They accepted English being spoken at home, although there was a recognition of the importance of being bilingual for cultural and economic reasons.

- **Children's agency:** The children described how they worked hard to fulfil parental expectations of 'being good', working hard at school and being respectful of their elders. At the same time, they were able to develop spaces for autonomous activity, to disagree with their parents and to draw on their own linguistic expertise to challenge parental intervention. Again, the cultural assumption of fixed and strict parent and child relationships was shown to be simplistic.

Evaluation

Like Shain's work, Lau-Layton's research serves to challenge dominant assumptions about minoritised communities in the UK, and reveals the ways that they are often 'othered' by the majority White discourse. Although her sample size was small, her repeat interviews over nine months meant that she was able to acquire a deep understanding of her respondents, which enhanced validity. Lau-Layton recognised the value of her own positionality to gain access to the research community, and she was careful to protect the children in the study.

Questions

What type of sampling did Lau-Layton use, and how representative do you think the sample was?

How does Lau-Layton's research design enhance or limit reliability and validity?

Question

What issues can you identify with Lau-Layton's decision to selectively transcribe the interviews she undertook?

Questions

Why was it important for Lau-Layton to undertake this research?

What insights does her study provide for the understanding of both education and British Chinese family life?

Interviews as a social research method

1. Considering that building rapport is essential to undertaking a good interview, how does rapport-building also impact the rigour of a research study?
2. Semi-structured interview design seems to be a popular choice for social researchers – why do you think this is the case? Draw on studies in this chapter to debate the strengths and weaknesses of different interview types.
3. To what extent can we generalise from interview data? Which study in this chapter had the most transferable data and why?
4. A number of interview studies in this chapter focus on family life and parenting. Why do you think interviews are a popular method to understand this topic? Would any other methods be more or less effective and why?

Advantages of interviews:

- They are said to have validity, because detailed and honest information is obtained when interviewer and interviewee develop good rapport. This is more likely in unstructured and semi-structured interviews.
- They can elicit people's close, subjective experiences.
- Unstructured and semi-structured interviews give people a chance to elaborate on their experiences through open-ended questioning, rather than being confined to closed answers.
- They disclose the ways that meanings are constructed rather than just facts.
- Interviews are more reliable when structured than unstructured or semi-structured as they are more easily repeatable.

Disadvantages of interviews:

- Interviewers can also express bias – they might ask leading questions, follow up, probe and prompt on points they are interested in, or give different cues (frowns, smiles, etc.). This may result in social desirability bias, where interviewees respond in the way that they think the interviewer wants them to respond, rather than with a truthful account.
- Unskilled interviewers can lead to poor data being gathered, particularly in an unstructured format.
- Interviews can be more time-consuming than other methods, such as surveys.
- Studies using interviews often have a smaller sample size than social surveys and so are less representative of the wider population. However, qualitative researchers may not be concerned about demographic representativeness, instead wanting to reflect the experiences of the people being studied.
- The use of open-ended rather than closed questions means that it is hard to make statistical inferences from the data. Generally, statistics derived from interview data are descriptive.

CHAPTER 6: OBSERVATIONS

What are observations?

While interviews provide the opportunity to hear the perspective of the people that you are researching, there is inevitably something artificial about the encounter. Therefore, social researchers can also choose to observe people as they go about their everyday business, in what might be described as their 'natural' settings. The home, workplace, supermarket, doctor's surgery, classroom, church or temple, gym, ballet class, restaurant, nightclub, etc. all provide spaces where a social researcher can undertake observations. These spaces are known as the **field** – the site where observation takes place. The term **fieldwork**, therefore, refers to the process of observing the lives of others and gathering data.

> **Key concepts**
> **field** a generic term that differentiates the natural setting from the laboratory; it defines the parameters of the research space; a domain of practice or action
> **fieldwork** the practical work undertaken by the researcher within the field; the process of immersing oneself in as many aspects of the lives of the people within a field

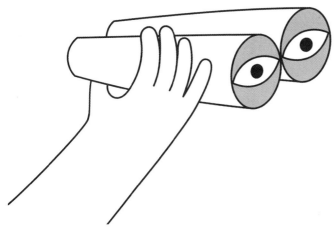

But how can observing people be a sociological method? We all engage in 'people watching' as part of everyday life, and it is often entertaining or thought-provoking to do so. Think about the occasions when you watched others. Perhaps you were waiting for a train or bus and observed the encounters of fellow passengers; perhaps you were interested to see which families got up first on holiday to secure their sun loungers by the pool; perhaps you noticed that a couple on the table next to you in a café had not spoken for the duration of their meal. These might be familiar observations – and you will have made many more.

However, observing as a sociologist is very different. It involves careful and systematic observation, documentation and checking, identifying patterns and taking account of (and describing) the wider context. For these reasons, observing as a sociologist is much more than people watching. To be a social scientific observer demands research skills and the ability to make a judgment about what sort of role the researcher should take during the observation. It also demands ethical care, so that the people being observed are protected, and reflexivity to ensure that the researcher's interests and bias are not dictating what they choose to see.

Observations on housework

Undertaking observations of what happens in someone's home can be very difficult, as it is not easy to get access. However, Szabo (2013) used observation as one of her methods for examining what happens when men cook at home. She recognised that women did most of the home cooking, so she wanted to see if men approached the 'chore' of cooking differently and what meanings they gave to the task.

Szabo took a constructionist approach. Alongside looking at diaries and undertaking interviews, she observed ten men (who all lived in the Greater Toronto area of Canada) as they cooked. She took extensive **field notes** to remember the kitchen space, the comments that the men made about the food, and the emotional descriptions of cooking that her respondents articulated or performed. She found that the men predominately saw food preparation as a 'leisure' activity: they took their time (it was leisurely!) and they listened to music or consumed alcohol while cooking and producing 'special' meals for others. Cooking, then, was an enjoyable activity for the men in this study.

Key concepts

field notes the qualitative notes taken by a researcher engaging in fieldwork; a record of observations

Question

How important do you think observations were for Szabo's study?

How do sociologists use observations?

What sort of observer can you be?

There are many different forms of observational enquiry. Choices need to be made about how involved the observer is in the activity that they are researching. Additionally, researchers need to decide the extent to which the subjects know that they are being observed. As such, there is a contrast between observational types that relate to:

- the mode of participation (whether it is more observational or more participatory)
- the degree of knowledgeability (whether the researcher is known or unknown to those being observed)
- the researcher positionality (whether the researcher has **insider** or **outsider** status).

The diagram below outlines these different approaches:

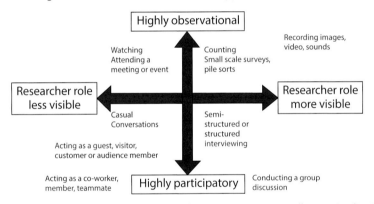

Figure 6.1: Approaches to research observations. *Source: Collecting Qualitative Data: A Field Manual for Applied Research* (Sage Research Methods).

Additionally, it is possible to distinguish between 'overt' and 'covert' presence. If the people you are observing know that you are a researcher, this is overt research. For example, if you were sitting at the back of a classroom and the students were told that a sociologist was joining the class that day, it would be overt **non-participant observation**. If the students were told that their teaching assistant was undertaking research, this would be a form of overt **participant observation**.

To be overt about your role as a researcher is, of course, very ethical. It means that everyone knows about the research, and you can provide opportunities for them to agree to or decline to be part of the study. However, a potential disadvantage of overt presence is that people may change their behaviour because they know that they are being studied/observed. One way to avoid this problem is to spend time

Key concepts

insider describing research that is undertaken within a group or community in which the researcher is already a member

outsider describing research that is undertaken by someone outside a group or community

non-participant observation when a researcher does not take part in any conversations or activity with the subjects (e.g. watching an interaction through a two-way mirror, listening to a conversation at another table in a café)

participant observation when a researcher participates in an event, conversations and activities

Question

Which of these types of observation would you consider to be the most ethical? Justify your answer.

> **Questions**
>
> What examples can you think of where people may change their behaviour if they know that they are being observed?
>
> What do you think the advantages and disadvantages of overt observation might be?

with the people that you are studying, building trust and rapport so that they forget that they are being researched and start to behave more naturally.

In contrast, covert research takes place when the people being researched do not know that they are being observed. This is almost like an undercover operation and comes with far more ethical dilemmas.

Observational research – like many other types of research – is also shaped by the positionality of the researcher. We can draw a crude contrast between insider and outsider status (also referred to as 'emic' and 'etic' perspectives). Insiderness, or an emic perspective, is said to provide the researcher with insight and access, and their presence in the field is less likely to cause people to behave differently (the Hawthorne Effect). However, there is a danger of bias, subjectivity and over-immersion. Outsiderness, or an etic perspective, is more detached and less biased, but it can be more difficult to 'pass' (to be unobtrusive) in the field and to secure trust from research participants.

 Becoming a social researcher

So far, we have identified four different roles an observer can take. Imagine that you want to research the gender division of cooking. Think of how you would undertake each type of observation in this setting and record your ideas in a copy of the table below. Alternatively, choose different research topics to explore how the various modes of observation would work. An example of covert non-participation has been included to get you started.

	Participant	Non-participant
Overt		
Covert		Observe different cooking and hosting rituals when visiting friends and family. Use a structured observation sheet to record the division of labour.

In practice, the role of the researcher can be more complicated than these four ideal types. In the first place, it is hard to be strict about the division between participant and non-participant observations – you may wish to be the latter but find yourself being caught up in the activity. Similarly, while you may endeavour to be overt about your role as a researcher, it is sometimes difficult to be sure that all participants are aware of your presence.

The most involved way to observe the lives of others is to fully immerse yourself in the field and spend as much time as possible with the people whose lives and experiences you wish to study, becoming part of their community. This form of participant observation is also sometimes referred to as ethnography, which literally means 'writing about cultures'. This method was traditionally associated with anthropology, but is widely used by sociologists and is often preferred as a term, as it encapsulates activities beyond simple observation. For instance, during ethnography, a researcher may decide to ask questions by using interviews and may collect documents or take photographs.

How do you get access to the field?

If you want to observe student behaviour in a library, it would be easy to organise if the students were working in a local library that anyone can use. However, the university library might be more difficult to gain access to – you might need to apply for a library card or ask someone to help you get into the building, for example. Gaining access to the social setting that you want to observe can be difficult, therefore, if you are interested in an activity that does not take place in public. Someone who helps you in these cases is known as a gatekeeper. A gatekeeper can be a friend, a colleague, an academic, a manager in an organisation or any other contact.

There are practical as well as ethical issues to navigate when trying to access private settings. For example, you might not be able to afford entry, or you might need to be a particular gender or age, sexuality or ethnicity to engage in observations in some settings. This is known as being able to 'pass' as a member of the group being studied, and it is especially relevant when the research is covert.

Key observational skills, tools and dispositions

Field notes

To be an excellent observer, you need to carefully document what you see. However, this can be difficult – especially if you are a covert observer, when writing down notes might give the game away. Many researchers in this situation hold on to their observations and then write up their diaries in the toilet, in the car or when they get home. Inevitably, there will be some loss of detail if you cannot record what is happening in real time, so it is important that field notes are made as soon as possible after the observations have taken place.

Thick description

One important characteristic of observation is to ensure that as much of the experience as possible is noted. An excellent observer will note not just the conversation, but the whole cultural setting: time, place, sounds, smells, clothing, and so forth. Geertz (1973) coined the term 'thick description' to capture the importance of observing all details, small and large.

Reflexivity

An excellent ethnographer also interrogates their own impact on the research, and will note this throughout the period of observation. This might include personal reflections in the field notes – if an observer can recognise when they feel anxious,

> **Becoming a social researcher**
>
> Getting into a university library might not be straightforward, but the access issues can be relatively easily resolved. However, if you want to observe private settings such as a religious cult, or a swimming class, or a private men's club, a gang, the police force, or an animal liberation group, for example, other access issues emerge.
>
> Choose three of these examples. Make a list under each example of the different access issues you might face when undertaking research.

annoyed, empathetic, etc., it can help them understand how their own values and dispositions were shaping what they saw and what they chose to note down.

Objectivity and detachment

It is important to remain an observer and not get too attached or too involved with the people you are studying. 'Going native' refers to those occasions when a researcher becomes so immersed in the research that they lose their objectivity, stop researching the setting and instead become a regular member of the group. However, it is worth considering the descriptor 'going native' with a critical eye. The term 'native' to describe this process maintains a colonial hierarchy that assumes the researcher is more objective and rational than the people they study. 'Over-immersion' is a less politically charged term that perhaps better explains the loss of researcher identity.

Legality and ethics

During ethnographic research, there is always a possibility of encountering behaviour that is illegal, dangerous or ethically problematic. Researchers must therefore take great care to mitigate these risks, and they will sometimes need to seek legal advice or disclose illegality. (See Venkatesh's study below for an example.)

Writing up research

When writing up an ethnographic study, the researcher relies on their interpretations, and makes decisions about what to include and how to represent what they observed. In other words, the act of writing is a construction of what happened, and is rooted in the researcher's own point of view. This is why the practices of reflexivity, objectivity and detachment are central to ethnographic practice. As such, researchers are usually transparent about their own positionality and how this shapes the observational encounter and their interpretations of what was 'going on'.

How rigorous are observations?

Observational research is often, like interview research, described as having validity but is less reliable. It is said to be valid because the researcher gains a powerful sense of why people behave in the way they do. However, there are two key issues of reliability in observational research.

Because the data is gathered from one person's point of view – that of the researcher – it cannot be said to be fully reliable. We must trust that the researcher's recordings and interpretations of events are true, as they are not corroborated by other research evidence. For example, in surveys, large sets of data can reveal patterns, and even in interviews themes can be discerned across multiple transcripts. Replicating observational research is also challenging, if not impossible.

Researcher insights

From 2018 to 2020, I (Sarah) carried out an ethnographic study in Margate, undertaking observations as well as interviews and focus groups. I accompanied local council workers as they conducted house visits and street audits (see Koch et al. 2020). In addition, the local community centre invited me to a Christmas party, and I joined the Roma community in an outside green space for an afternoon of music and food.

During my fieldwork, I visited homes and saw first-hand the standard of some of the rental properties, kept clean despite the overcrowding, damp, poor décor and loose electrical fittings. In contrast, I was also welcomed into five-storey houses, recently bought and renovated by creative folk who had moved to the town in search of cheap property and artistic stimulus. I spent much time observing the day-to-day comings and goings in local cafés, pubs and galleries.

The various observations involved taking different roles. Accompanying public-sector workers as they went about their daily work situated me in a covert non-participant role. In contrast, when attending parties, my presence was overt and took the role of participant. I engaged in face painting and making decorations with the children. When hanging out in the cafés and pubs, I was again situated in a covert role, but I was open with people when I approached them for an interview. I was often viewed with suspicion – regarded as a journalist looking for a scoop. Having university ID and consent forms was necessary then, not just for ethical reasons but also to secure access and trust.

Undertaking observations sometimes felt dangerous, especially when I was out alone at night. It was also extraordinarily time-consuming and sometimes uncomfortable, especially when I witnessed racist conversations that I found personally abhorrent but needed to hear in order to gain a comprehensive understanding of people's values and experiences. Working out how to present this data proved to be a significant challenge.

The study described how the parallel universes of the creative elite, the White working class and working-class migrant communities rarely connected. There was evidence of tension and mistrust between and within each main group. At the same time, there was evidence of conviviality with shared community projects. The nuance of what it meant to share a community was dramatically revealed, providing a case study of the ways in which social inequalities in the UK are mapped and diversely experienced.

Which sociologists are likely to use observations?

Observational research is used by different sociologists, ranging from positivist and realist approaches to interpretivist and constructionist. The approach the researchers take will differ. Positivists are more likely to use structured and systematic observations, such as Humphrey's observation sheets (see below), while interpretivists may rely on reconstructed evidence as told from the perspective of the researcher. This is sometimes narrated as a 'story' (see Venkatesh's account of the time he spent with a Chicago gang, below).

CLASSIC RESEARCH: SEX IN PUBLIC PLACES

Humphreys, L. (1975). *Tearoom Trade: Impersonal Sex in Public Places*. New Jersey: Transaction Publishers.

Study overview

> At shortly after 5 o'clock on a weekday evening, four men enter a public restroom in the city park. One wears a well-tailored business suit; another wears tennis shoes, shorts and tee-shirt; the third man is still clad in the khaki uniform of his filling station; the last, a salesman, has loosened his tie and left his sports coat in the car [...] What common interest has brought these men to this public facility? They have not come here for the obvious reason, but in search of 'instant sex'. (p. 29)

The men that Humphreys describes in the quotation above had met at 'tearooms', as they were referred to in the gay **subculture**. His book *Tearoom Trade* describes his two-year-long study observing homosexual activity in men's public toilets in the United States.

Method and sample

Humphreys used participant observations and interviews to gather data on sexual encounters between men. At the time of his study, these acts were considered socially deviant and therefore had been under-researched.

To gain access to the field, Humphreys thought that he had to fit in (pass) so as not to disturb the 'normal' goings-on. He therefore decided to undertake covert research by 'being another gay guy' (p. 24). In order to effectively observe behavioural patterns and to record data, he adopted the role of the 'voyeur' (look-out) – known as a 'watchqueen' (a gay man who gains pleasure from watching other men participate in public sex, but also warns them if the police are nearby).

Humphreys notes the challenges of undertaking covert research:

> I had to enter the subculture as would any newcomer and to make contact with respondents under the guise of being another gay guy. [...] Such entry is not difficult to accomplish. Almost any taxi driver can tell a customer where to find a gay bar. A guide to such gathering places may be purchased for five dollars. The real problem is not one of making contact with the subculture but with making the contact 'stick'. Acceptance does not come easy, and it is extremely difficult to move beyond superficial contact in public places to acceptance by the group and invitations to private and semi-private parties. (p. 24)

Humphreys split his research into two main phases. Phase 1 involved unsystematic observations, which he recorded retrospectively with a tape recorder in his car. This gave him some general understanding of the tearoom culture. Phase 2 involved more systematic observations in order to achieve objective validity (to ensure the

Questions

Why do you think Humphreys had to be accepted into the subculture?

What are some of the challenges that he faced?

data was not influenced by his own presence or presuppositions) and reliability (to ensure that his study could be replicated). He devised an observation sheet, which he completed for 50 of the sexual encounters he observed. A further 30 encounters were also observed and recorded on these sheets by a 'co-operating participant'.

Within this observation sheet, he recorded:
- who was present
- their position in the room
- their clothing
- their age
- the time of day.

Humphreys revealed his identity to 12 participants, with whom he developed trusting relationships. He interviewed them as a fact-checking exercise to corroborate his observational data. He then undertook 50 further formal interviews with men he had observed. However, these men were sampled very differently to his convenience sample of 12.

During his observations, Humphreys noted down the men's car registrations. He used these to find out the men's names and addresses by approaching 'friendly policemen', who gave him access to the licence registers without asking why he wanted them. Once Humphreys had gained this information, he observed the streets in which the men lived, and learned about their family life, including whether they had children (inferred if there were swings or slides in their garden, for instance).

Humphreys then went one step further. He was involved in another piece of survey research using structured interview schedules into the subject of men's health. He arranged for 100 of the car owners' details to be included in this broader survey (what he called the 'deviant sample'). He and a graduate student, who was entrusted with knowledge of his tearoom trade study, interviewed these men face to face under the guise of the social health survey.

Humphreys waited a year before he conducted these interviews, and protected his identity by changing his appearance, clothing and car. He reported that he recognised the men, but he did not think they recognised him.

After the survey was completed, Humphreys took out the results of this sample for which only 50 had been fully completed. He used the data gathered from the questionnaire to gain some basic demographic information about the men who were engaging in tearoom activities, such as their age, marital status, etc.

Findings

Humphreys' main findings were that:
- most men participating in tearoom activities were married and living with their wives and children
- the men did not pose a threat to anyone else – no one was approached in the tearooms, propositioned for sex or recruited against their will.

> **Question**
>
> What does Humphreys' use of an observation sheet tell us about his theoretical orientation towards social research?

> **Question**
>
> What ethical issues does Humphreys' mode of access to interviewees raise?

> **Question**
>
> What problems, if any, can you identify with the inferences that Humphreys makes here?

At the time of the study, prevailing stereotypes about gay men were that sex in toilets was an activity exclusive to gay men, and that it was often predatory. Humphreys' findings challenged the stereotypes. His results showed that out of 50 tearoom participants interviewed:

- 54% were currently married
- 8% were divorced or separated
- 14% were openly homosexual
- 24% were single and not openly gay.

The men involved in tearooms, Humphreys claimed, tried to mitigate their sexual activities and took on a defensive 'shield of superpropriety' to ward off social disapproval (p. 135). Concerned about exposure and stigmatisation, Humphreys argued that the men overtly presented themselves as morally upstanding, through having neat homes, driving new and clean cars, being well groomed and wearing good clothes, devoting time to family life, and being involved in religion.

Evaluation

Humphreys' research raises several ethical issues. He specifically identified, and tried to mitigate, three key ethical concerns with his own study:

- **Deception:** Humphreys pretended to be someone else to gain information. In his study he claimed that, because he took on the role of 'watchqueen' he was not misrepresenting himself – he *was* a voyeur, albeit in the sociological rather than the sexual sense. He also argued that he was observing a public place (a toilet) rather than a private domain. With regards to the social survey aspect of his research, he claimed that it was common practice to gather data about people without respondents knowing the full extent of such data gathering.
- **Confidentiality** and **anonymity:** There was a risk of the research participants' identities being exposed. Humphreys stated that he was incredibly careful to ensure that no one's identity was revealed.
- **Harm:** The research could have a negative impact, for instance, on the population being observed. Humphreys argued that his research was incredibly important for wider society.

> **Key concepts**
>
> **confidentiality** when data is modified to ensure that respondents cannot be identified from data (e.g. by using false names)
>
> **anonymity** when data is collected without obtaining personal or identifying information from participants
>
> **harm** physical or psychological damage that can result from a research study; a researcher has a responsibility to protect participants as well as themselves from harms, and often carry out risk assessments to minimise harms

> **Questions**
>
> Do you think that Humphreys successfully mitigated these three ethical issues? Why, or why not?
>
> Do you think it would be possible to replicate this research today?

Humphreys argued that while his research raised ethical concerns, all research methods entail some ethical issues:

> Criminologists may study arrest statistics [...] without stirring from the safety of their study chairs, but such research methods may result in the creation of a fictitious 'crime wave', a tide of public reaction, and the eventual production of a police state – all because the methods may distort reality. There are no 'good' or 'bad' methods only 'better' or 'worse' ones. (p. 169)

Question

Do you agree or disagree with Humphreys?

You can apply the STRIVE rules to Humphreys' study.

- **Sociological:** The work is clearly of sociological significance. During the time of Humphreys' research, sex between men was heavily socially stigmatised and therefore under-researched.
- **Theoretical:** Humphreys took a positivist approach to observations, using observation sheets to omit bias and to avoid error in information recall that is sometimes found in ethnographic research that otherwise relies on memory.
- **Representative:** Although his research aimed to be scientific, it is hard to claim it was widely representative, as it was based on a small sample of men in a particular area of the USA.
- **Impactful:** Humphreys argued that his research was impactful because it aimed to challenge common stereotypes about gay men and highlighted that men participating in illicit sex in 'tearooms' were otherwise 'ordinary' – not fitting a particular 'type'. For example, they were often married with children, and came from a range of backgrounds. He also suggested that his research was impactful for debunking myths that gay men are predatory towards younger people. During the tearoom encounters, no heterosexual men were approached, and sex only took place between consenting men.
- **Valid and reliable:** The study has validity because it directly observed behaviour in a natural setting, but it is not necessarily reliable because the study is challenging to repeat.
- **Ethical and reflexive:** It is unlikely that the study could be carried out today, as its covert nature presents too many ethical challenges, particularly concerning deception and lack of informed consent.

You could also apply the qualitative judgment criteria from Chapter 1 to Humphreys' study.

CONTEMPORARY RESEARCH: CRIME: GANG LEADER FOR A DAY

Venkatesh, S. (2008). *Gang Leader for a Day*. London: Penguin Press.

Study overview

> The young men rushed up to me, within inches of my face [...] someone asked what I was doing there. I told them the numbers of the apartments I was looking for. They told me that no one lived in the building. Suddenly some more people showed up, a few of them older than the teenagers. One of them, a man about my age with an oversize baseball cap, grabbed my clipboard and asked what I was doing. I tried to explain but he didn't seem interested... 'Julio over here says he's a student,' he told everyone. His tone indicated that he didn't believe me. 'Who do you represent?' [...] 'Represent?' I asked. [...] 'We know you're with somebody, just tell us who.' [...] Another one, laughing, pulled something out of his waistband. At first, I couldn't tell what it was, but then it caught a glint of light and I could see that it was a gun. (p. 13)

Venkatesh began his research into gangs in 1996, but this was underpinned by his experience as a graduate student at the University of Chicago in 1989. As depicted in the excerpt above, as a graduate student he had visited a housing project in Chicago to undertake a questionnaire (as part of some research he was working on for one of his professors) with people residing in The Projects (public housing in the USA). In his words, he wanted to understand what it was like to be Black and poor.

Method and sample

In the scene described, Venkatesh explained that his initial encounter with the Black Kings, a local gang, also forced him to reflect on the research methods he would later come to use in his own research. The Kings did not believe that he was there to undertake survey research. Instead, they thought he was a rival gang member, so they kidnapped him and held him for the night. They searched his bag and found his survey and clipboard. Handing it back to him, they told him to go ahead and ask a question. Sweating and nervous, Venkatesh describes the encounter further (swear words redacted):

> The first question was one I had adapted from several other similar surveys; it was one of a set of questions that targeted young people's self-perceptions.
>
> 'How does it feel to be black and poor?' I read. Then I gave the multiple-choice answers: 'Very bad, somewhat bad, neither bad nor good, somewhat good, very good.'
>
> The guy with the too-big hat began to laugh, which prompted others to start giggling... 'You got to be [...] kidding me.' (p. 14)

During his initial encounter with the Black Kings, Venkatesh met JT, the local gang leader. Venkatesh describes being held overnight, recounting that JT came the next morning after the gang had discussed what to do with him, having realised that he was not another gang informant.

Questions

What does the description of this encounter reveal about the use of survey methods to understand certain sociological questions?

Why do you think that, through his initial encounter, Venkatesh realised that survey methods would not elicit the information he needed, and that he would have to instead immerse himself in the field to gain these insights?

On letting him go, JT advised:

You shouldn't go around asking them silly-ass questions. [...] With people like us, you should hang out, get to know what they do. How they do it. No one is going to answer questions like that. You need to understand how young people live on the streets. (p. 21)

Venkatesh took this as an opportunistic invitation to return, and seven years later he finished conducting his observational research. His relationship with JT, who later took him under his wing and 'sponsored' him, provided Venkatesh with access to a gatekeeper who would enable him to conduct research in the field with gang members and learn about life in The Projects. While this gave him access to the Black Kings and to some other influential inhabitants of The Projects, JT's position in the gang's hierarchy limited his access to individuals higher up in the gang organisation.

Venkatesh chose ethnography because of the limitations of scientific modes of inquiry into social problems. For example, he showed a mismatch between what the official statistics suggested and what he found on the ground when observing The Projects:

While the official statistics said that 96 percent of [...] adult population was unemployed, many tenants did have part-time legitimate jobs – as restaurant workers, cabdrivers, cleaning ladies in downtown corporate offices, and nannies to middle-class families. But nearly all of them tried to hide any legitimate income from the CHA, lest they lose their lease or other welfare benefits. (p. 195)

In his research, Venkatesh also conducted interviews with people involved in other aspects of life in The Projects – food, sex work, etc. Venkatesh shared his findings with JT and other gang members, who used them to punish anyone who earned money but did not pay their 'taxes' to the gang. Venkatesh therefore inadvertently placed his research subjects in danger. You can see from this example that protecting respondents is an on-going and reflexive responsibility.

Venkatesh himself had to seek legal counsel during the study because he had knowledge of criminal activity (he knew of a planned drive-by shooting). He was informed of his obligation to report any planned crimes to the police. In his book, he discusses his concerns and the challenges of disclosing such legal obligations to close associates in the gang like JT.

Questions

Is there any evidence to suggest that Venkatesh became so immersed in the field that he was unable to step back from the study with impartiality (over-immersion)?

Do you think he had any reason to feel guilty? Why, or why not?

Why do you think leaving the research setting and respondents might be more challenging with ethnographic research than with other research methods?

Questions

What does this excerpt suggest about the difference between using official statistics as methods and the use of ethnographic methods?

Which method do you think is more useful for learning about unemployment and why?

Towards the end of his research study, Venkatesh reported on his experience of exiting the field. He had formed deep and lasting bonds with JT and other informants in his study. When he left the field, The Projects were being torn down and some of the informants had been arrested. While JT remained out of prison, his gang was crumbling. Venkatesh described his guilt about leaving and spending less and less time with the Chicago gangsters.

Unlike Humphreys, Venkatesh did not disguise his identity as a social researcher – to this end, his research was overt. He did, however, still have to 'pass' with the people he wanted to research. He reflected on his positionality, suggesting that being South Asian meant that he fitted in better than if he had been of White ethnicity, particularly as the White people populating the neighbourhoods he was observing tended to be police.

Case study: *On the Run* and positionality

Venkatesh's South Asian ethnic identity enabled him to fit ('pass') in his ethnographic field better than if he had been White. In her study *On the Run: Fugitive Life in an American City*, White researcher Alice Goffman also undertook ethnographic field research with Black men in poor neighbourhoods, this time in Philadelphia. She wanted to understand about how poor Black men were criminalised and treated by the criminal justice system. Parry (2019) openly challenged Goffman, asking: 'Should privileged, White outsiders tell the stories of poor minority communities?' This again highlights the importance of reflexivity, which is bound up in Bejarano et al.'s (2019) call to acknowledge White privilege and the hierarchies of power and inequality that sociologists may reproduce.

Questions

How do you think Goffman's positionality might have affected her ability to 'fit in'?

Are there any ethical issues regarding her positionality in the study? If so, what are they?

How might other socio-demographic characteristics shape the relationship between the researcher and their participants?

Questions

Venkatesh described himself as a 'rogue sociologist, breaking conventions and flouting the rules' (p. 283). Why do you think he describes himself in this way?

Do you consider his ethnographic work 'rogue' compared to a) research using other methods and b) other ethnographic pieces of research?

Questions

In his author's note, Venkatesh wrote: *Wherever possible, I based the material on written field notes. Some of the stories, however, have been reconstructed from memory. While memory isn't a perfect substitute for notes, I have tried my best to reproduce conversations and events as faithfully as possible.* (p. 285)

Reflect on this excerpt – what does this suggest about the study's reliability? How does it compare to Humphreys' approach?

Findings

The key aim of Venkatesh's study was to provide a more nuanced understanding of the behaviours and worldviews of those who were Black, poor and members of gangs. Significantly, he revealed that gangs, which were often presented as being in opposition to local communities, helped many aspects of community life and were sometimes seen favourably by poor community members. For instance, because The Projects were seen as a no-go zone by police, gangs kept 'order' and people felt that things could be worse without the gangs.

Evaluation

Overall, Venkatesh's interpretivist study provides deep insight into the hidden and inaccessible life of gangs, which is rarely subject to sociological analysis. His output is story-like rather than scientific, as he acknowledges. The rich insights gleaned from his ethnography are nevertheless deemed so important that the compromises that ethnography poses to validity, reliability and representativeness are acceptable. Venkatesh is upfront about how his own positionality secured him access to the field.

It is also possible to make several inferences about the ethical implications of ethnographic methods from Venkatesh's study:

- Ethnographic research is potentially dangerous to researchers.
- Ethnographic research can be harmful to those being researched.
- The people social researchers want to study may be wary of their presence in the field.

> **Question**
>
> Reflect on the different harms that the ethnographic research might have posed to the communities being studied in Humphreys' and Venkatesh's studies. Which do you think was more harmful and why?

Children outside a Housing Project on Chicago's South Side

CONTEMPORARY RESEARCH: FAMILY: ARRANGED MARRIAGE

Pande. R. (2021). *Learning to Love: Arranged Marriages and the British Indian Diaspora*. New Jersey: Rutgers University Press.

Study overview

'I think that you should have a chapter in your book describing in detail how we arrange marriages and how intricate and wonderfully cultural a phenomenon it is. Not like in the West where you just produce a person from somewhere and tell your parents that this is who you plan to marry.' (p. 42)

Pande's intergenerational ethnographic study of arranged marriages, as they are practised among the British Indian **diaspora**, was designed to explore the complexity of marital practices within the community. She wanted to challenge prevailing stereotypes held in western media and immigration policy about familial decisions in marital choices, where arranged marriage is viewed with suspicion and seen as 'cold practice' rather than 'warm love' (p. 20).

Method and sample

Pande described her method as 'ethnographic' rather than ethnography, because she was not fully immersed in the field. Her methods involved interviewing 44 respondents, and she supplemented all interviews with extensive field notes. She argued that ethnographic methods were necessary to capture the nuance and complexity of arranged marriage in the British Indian diaspora.

In particular, she focused on showing how western discourses often conflate arranged marriage with forced marriage. In doing so, she felt that they failed to understand the intricate and careful decision-making associated with marriage *choices*. Pande showed that arranged marriage takes on many different formations, which she describes as an 'elastic practice' (p. 5) rather than an enforced rule.

Pande importantly used decolonial methods in her study, as she wanted to reveal how dominant discourses about arranged marriages are shaped by colonialism. Arranged marriage was often viewed in negative terms – in some way deficient when compared to 'western' marriage decisions. 'Western marriages' were often framed as being rooted in 'falling in love'. Overall, dominant, colonial cultural assumptions associate love with choice, while arranged marriages are associated with coercion.

> ### 🧠 Thinking further
>
> Despite Pande describing her work as ethnographic, her research method is significantly different to the approaches taken by the other ethnographic studies described in this chapter (Humphreys, Venkatesh and Garthwaite). While most ethnographers use interview methods to provide data alongside their observational work, Pande did not immerse herself in the everyday lives of those she studied. Instead, she supplemented interview data with notes about the setting and social context in which the interviews took place. For example, she described sharing meals with respondents, as well as sitting through prayer services or talking with community leaders while waiting to interview respondents. These experiences documented in her field notes were integrated into her analysis.

Pande's research included three broad groups of participants:

- First-generation migrants who had (mainly) come to the UK in the 1960s. They had an arranged marriage forged in India, with one spouse following the other to the UK. This group included corner-shop owners, property developers, lawyers, doctors, academics, tailors and restaurateurs. They had arranged their children's marriages or intended to do so.
- First-generation migrants who had come to the UK in the 1990s. Most of these participants went back to India to find a wife or husband. A small number married British-born Indians. Members of this group were mainly employed as IT professionals, engineers and academics. They had also arranged their children's marriages or intended to do so.
- Second-generation migrants who had often been born or raised in the UK. Among this group, marriages had predominantly been arranged, or members were looking for a marriage at the time. Here, occupations included corner-shop owners, property developers, students, restaurateurs, waiters, doctors, engineers and teachers.

Most of the sample identified as Hindu (60%); the remainder identified as Sikh (30%) and Indian Muslim (10%).

Pande's sample enabled her to look at intergenerational change as well as the influence of religion, migration and social class. She spoke English, Hindi, Urdu and Punjabi, and was able to talk to respondents in their first language. For the purposes of her research, she had to translate all the responses into English. However, she wanted to keep certain literal (if unfamiliar) translations – Indian-English **neologisms** – to capture the nuanced practice of arranged marriage. Pande and her respondents noted that certain words and ideas cannot be easily or comprehensively translated, so to explain the nuanced processes of writing culture, she included some of her deliberations of translation. In the excerpt below, you can see how the word 'love' is used differently across cultures, acknowledged by her interviewee, Mrs Pal:

The word love in English rolls of the tongue so easily, I love drinking tea, I love London and I love you, you speak Hindi don't you? Try translating these sentences… It can't be done. At least I can't do it, can you. Without making it sound trite and insincere. How can the same emotion be used to express my love for food and a person? Just doesn't work. (p. 87)

Question

Do you think ethnography is a social scientific method? Make your case for or against this.

 Thinking further

In describing the importance of language and the translation of respondents' accounts, Pande draws attention to the difficulties of writing up ethnographic research. In the process of writing about culture, an ethnographer has to decide what to report, how to frame observations, and how to represent what they see. While the writing up of an ethnography, or indeed any research, involves choices about what to include, this does not mean that writing is entirely subjective and lacks rigour. However, it is important for the author to recognise their own positionality and how this shapes what is seen, heard and represented through their words.

Findings

Pande's research revealed that the term 'arranged marriage' actually includes a spectrum of marriage practices, and that it is central to the construction of identity, creating links to the 'home country' and passing on and preserving culture. Four main types of arranged marriage were identified, each revealing careful and participatory modes of decision-making and opportunity for choice.

1. Traditional arranged marriages, characteristic of the first-generation migrants, where parents and close relatives take the lead in choosing a partner. The men and women in this arrangement agree to meet a suitor by looking at photographs. This is followed by family meals, although contact between the couple is never permitted alone and is often minimal ahead of the wedding.

2. Semi- or partly arranged marriages, where a suitable match is decided through consultation with the couple and there are opportunities to get to know one another and 'fall in love' before the wedding. In other words, the run-up to the wedding involves courtship; all the people Pande spoke to said they had been 'in love' by the time the marriage took place.

3. Love-cum-arranged marriages, characteristic of the second generation, in which individuals are free to choose their own spouse and then the parents take over the arrangements and agree the match. In these cases, respondents talked about the importance of making sure they fell in love with the 'right' sort of person – someone whom their parents would approve of.

4. Arranged wedding, describing situations in which a couple independently fall in love and then parents take charge of organising the wedding ceremony. This usually includes traditional Indian rituals, even if the chosen partner comes from a different ethnic background. Parents approve of and pay for the wedding.

In all types of arranged marriage, Pande highlights choice but also an enduring commitment to parental involvement. In this way, the respondents felt their marriage choices were distinct from the 'incidental' nature of western marriages (p. 113). Romance was part of the decision-making, but emphasis was placed on loving the 'right' person and 'learning to love' rather than 'falling in love'. Such foundations were deemed to secure a stronger, longer-lasting marriage and, importantly, were still characterised by love. Pande's work reveals how romantic love is woven into arranged marriages rather than being distinct from them.

Thinking further

You might want to compare postmodern discussions of 'western liquid love' with Pande's account of 'learning to love' within arranged marriage. Bauman coined the term 'liquid love' to describe the impact of **individualisation** on relationships. He noted that bonds between couples had become more fragile in postmodern times because people desire more freedom at the same time as seeking security in relationships.

Pande's research also shows that arranged marriages are not 'fixed' and prescribed as they are so often stereotyped. Consider the importance of:

- recognising diverse family types
- challenging 'westernised' constructions of love and marriage
- challenging the 'othering' and stereotyping of non-western approaches to love and marriage.

Evaluation

Pande reveals how much sociological research looks at the world through a western colonial lens and western constructs of what counts as 'love'. Her research challenges stereotypes around British Indian diasporic culture and specifically about arranged marriages, revealing that familial involvement in marriage is varied and does not preclude agency, choice or love. In evaluating this study, it is worth referring to Pande's own summary of her approach:

> In order to genuinely appreciate the multi-ethnic nature of British society one needs to better understand the minority populations' own discourses of their cultures and traditions. Instead of using the yardstick of **hegemonic** norms to manage and control minority cultural practices, one needs to regard them as part of Britain, as another thread, among many, woven onto the tapestry of **postcolonial** British society. (p. 118)

Her study provides a deep, nuanced and valid insight, and her varied sample increases her claims of representativeness.

CONTEMPORARY RESEARCH: EDUCATION AND INEQUALITY: DEBUNKING CULTURAL DEPRIVATION

Brown, C. (2015). *Educational Binds of Poverty. The lives of school children.* London: Routledge.

Study overview

A body of research into the underachievement of certain groups of children in school points to how cultural deprivation – poverty and cultural values – shapes educational outcomes. This perspective is imbued with moral assumptions, suggesting that the poor are somehow responsible for their 'failings'. Bad families and bad attitudes are blamed for educational underperformance. These sets of assumptions have led to policy initiatives such as fining parents if children miss school.

Cultural deprivation is a difficult concept for many sociologists. While there may be correlations between poverty and educational achievement, and while research studies can establish links between holding certain dispositions and values and school engagement, it is too simplistic to see this as a causal relationship.

Perhaps, then, something else sits behind this correlation that can better explain educational attainment while avoiding the family-blaming that cultural deprivation theories seem to encourage.

By undertaking a longitudinal ethnographic study of six children who were situated in poverty, Brown wanted to understand what school life was like for them. She aimed to discover how the children managed their own lack of economic resources, navigated school culture, interacted with friends, fared in terms of teacher expectations, and steered their way through education while moving between schools.

Method and sample

Brown followed the children as they moved from Years 5 and 6 in primary school to Years 7 and 8 in secondary school. She conducted interviews with the children as well as observing them in the classroom. Brown was fortunate to already have contact with 50 schools because of a previous research project. She purposively sampled three of these schools because they were characterised by significant student mobility, whereby families were frequently moving in and out of the local area (what Brown calls 'turbulence'). One school was near an RAF site, another near a Traveller site and the third near social housing. Of these three schools, one was identified as having a more middle-class catchment (officers from the RAF base) and pupils at the other two schools were identified as being of lower social class. Brown selected 17 children from these schools based on their 'turbulent' educational trajectories – five were from the officer class and 12 from low income.

> **Activity**
>
> In pairs, discuss the difference between correlation and causation. Refer back to Chapter 1 to check your understanding.

Despite beginning with 17 students, Brown succeeded in following only six students. This was due to the challenges of longitudinal research, particularly when studying mobile populations: many of the students moved away so could not be followed up.

Brown recognised that because her sample size was small, her work could not be considered representative. Nevertheless, it was important in understanding children's accounts of how turbulence affected their learning, friendships and educational wellbeing.

Each chapter in Brown's book discusses a child in detail, providing intricate details of home life and school experiences:

1. Clive, who negotiated his education alongside relocating school as his family became headed by his mother alone.
2. Liza, who relocated to live with her father after the breakup of her parents' relationship.
3. Megan, a highly mobile Irish Traveller.
4. Codie, who had to move schools as the family could not afford their housing.
5. Helen, who relocated school following bullying.
6. Robin, who had a hearing impediment, and relocated school following parental redundancy.

> **Questions**
>
> What is the value of following six children for an extended period?
>
> What insights do you think this research generated that surveys, official statistics or interviews could not reveal?
>
> What questions would you ask about the study's ethics and representativeness?

Findings

In all cases, Brown found that the children had to navigate difficult circumstances – what she referred to as the 'binds of poverty'. The lack of material resources available to them made it difficult for the children to feel part of school life. Moreover, school felt alien to many of them because they moved between institutions so frequently. This was especially true for Megan.

The children often found it difficult to make friends, but when they did secure friendships, they described positive experiences of school. Finally, having to move schools was shown to be highly disruptive to learning and a sense of belonging.

The ethnography revealed the hard work the children undertook to simply get by in school. Brown describes their resilience as 'remarkable' (p. 164). Thus, rather than blaming and denigrating underachieving children and their families, she argued that there should be more help to support children to build friendships with their peers and relationships with teachers, as well as more financial support.

Evaluation

Brown's research is highly reflexive, in that she unsettles a longstanding idea in sociology – cultural blaming. She shows that structures of poverty have a deep impact on the personal troubles that students experience in school, but she also demonstrates how students are not without agency, and the remarkable tenacity and resilience that also characterises this population.

The sample size was very small, and therefore demographic representativeness was not possible, but it was not intended to be so. However, the research is highly valid as it focuses deeply on the children's school lives. Brown gets to know the students through her longitudinal ethnographic research in ways that other research methods, or even time-limited ethnographies, may not enable.

CONTEMPORARY RESEARCH: INEQUALITY: 'FOODBANK BRITAIN'

Garthwaite, K. (2016). *Hunger Pains: Life inside foodbank Britain*. Bristol: Policy Press.

Study overview

> *Maureen reflected on a recent experience of a woman she'd met in the foodbank who could have got a voucher, but would not ask for one:*
>
> *She came in with this woman and she could have had a voucher as well but she said 'I couldn't get on because I was just too embarrassed about getting it,' and it is hard but I said to her 'It's no good being embarrassed if you can't eat.' You know, I mean there's no need to be embarrassed because you are entitled to it and you're in need, and we're happy to do it for you. But at the end of the day you can't stop people feeling that way about it. (p. 146)*

Garthwaite's study focused on how reliance on foodbanks is stigmatised in Britain in a context of ever-growing social inequality.

Method and sample

Garthwaite engaged in a five-year study, of which 18 months was spent immersed in the field, volunteering every Friday and some Wednesdays at a local Trussell Trust foodbank. She also spent ten months at the Citizens Advice Bureau (CAB), which distributed red vouchers (for foodbanks), and took part in twice-yearly supermarket collections at Tesco, asking people to contribute to the foodbank.

She focused her research study in Stockton-on-Tees, where she was residing. This is the 27th most deprived local authority in the UK, and an area characterised by significant health inequalities. For example, Garthwaite noted that a man living in the most deprived ward in Stockton-on-Tees had a 17.5-year lower life expectancy than a man living in the most affluent ward (p. 18).

Garthwaite made two key sampling choices. Firstly, this was sampling out of convenience, as it was where she lived. Secondly, it was purposive – she purposefully chose an area of deprivation where foodbanks were more likely to be prevalent.

The fieldwork was divided between Garthwaite's volunteer work at a foodbank and time spent in one of the least deprived areas of the borough attending coffee mornings, visiting delicatessens, churches, mother-and-toddler groups, a multiple sclerosis group, yoga classes and a credit union (a type of money-lending institution).

> *I wanted to find out what it was really like living in Stockton-on-Tees, an area with such stark health inequalities, where jobs are minimum wage, zero hours and hard to come by, and where affluence and poverty exist side by side. (p. 17)*

Her ethnographic research comprised a series of field notes that documented her experiences in the foodbank, including notes on conversations (unstructured

---

Question

What are the advantages and limitations of Garthwaite's sample?

Questions

Why do you think Garthwaite spent time in places such as yoga classes and delicatessens, alongside time in a foodbank? What assumptions is she making about social class in these choices?

interviews) she had with foodbank users as well as volunteers and those who referred people to foodbanks. In total, Garthwaite talked to 80 people: 60 foodbank users, 12 volunteers and eight staff at foodbank referral agencies.

Most of Garthwaite's conversations took place in the foodbank where she volunteered, and she recorded these in her field notes. She claims that this methodological decision was due to the 'sensitive nature of the encounter' (p. 31). In this her field notes relied on notes and memory. However, she conducted 10 more conventional research interviews in people's homes, which were recorded. She gave £10 Love to Shop gift vouchers to these participants in exchange for their time.

Garthwaite conducted her observational research overtly, ensuring that everyone she spoke to was aware of her presence as a researcher. She omitted anyone from her study whose information she captured in her field notes, but to whom she was unable to disclose her researcher status. Garthwaite reflected on having to balance her simultaneous roles as volunteer and researcher. She noted that sometimes it would have been inappropriate and unethical to conduct her research, and she would step back into her role as volunteer, particularly if she was comforting someone who was upset. In this regard, she wore two hats and had to make judgment calls to ensure that participation in her study was ethically sound. Overall, she was committed to ensuring that her research was always voluntary, informed and confidential.

> **Activity**
>
> What can you say about the reliability, validity and ethics of Garthwaite's two types of interview data? Create a table to record your ideas.

Findings

Garthwaite's work involved debunking many myths about foodbank users that were perpetuated through the government and popular media. She presented this data in her book as 'foodbank myths vs. real life'. These myths include:

- poor people smoke
- poor people can't cook
- poor people own big TVs, take drugs and own well-kept dogs.

Garthwaite's ethnographic research gave space for reflection of foodbank users on these contentious topics. She also drew on social policy and government reports to reveal how foodbank users were represented – for example: 'A family earning £21,000 a year [...] where both parents smoke 20 cigarettes a day will spend a quarter of their income on tobacco' (p. 65).

Garthwaite noted that there may be a reality to these government descriptions, in that many poor people do smoke (like the wider population) and that spending money on cigarettes reduces the amount available to spend on food. However, in juxtaposition to these stigmatising and shaming criticisms, Garthwaite also noted that people on benefits using foodbanks are expected to give up small luxuries. Respondents often claimed that smoking was 'their only habit' and that it enabled them to cope with their stressful lives (p. 65).

Evaluation

Garthwaite's research focuses on a core sociological issue – that of poverty and social inequality. Her work also meets the STRIVE rule of impact – indeed, she describes the British Sociological Association's report on an event on food poverty in 2015 that detailed the importance of research impact:

> *Research and policy making needs to be rooted in real life with people's experiences at the forefront, so, researchers also need to explore how they can work more fully alongside people experiencing poverty and food insecurity. (p. 29)*

While the study focused on only one foodbank in Britain, the length of time she spent observing foodbanks and the large number of in-depth interviews she undertook enabled Garthwaite to give a highly valid account of foodbanks (although of course observations also rely on memory, which can compromise validity). Like other ethnographies, Garthwaite's study might be criticised in terms of representativeness and reliability. Since the study focuses on only one foodbank, it is hard to say whether it is representative of other foodbanks across Britain. Likewise, it is hard to say whether other people repeating this study or carrying out a similar study elsewhere would share Garthwaite's findings and draw the same conclusions. However, considering broader sociological knowledge about issues of poverty, inequality and stigma, Garthwaite's work would seem to have a wider resonance.

Observations as a social research method

1. Why might social researchers choose to use observational methods rather than interviews to understand social phenomena? Draw on studies from this chapter to illustrate your answer.

2. What insights and challenges do different degrees of participation (non-participant or participant) and presence (covert or overt) afford social researchers? Use the studies in this chapter to identify different types to illustrate your answer.

3. Use the studies in this chapter to reflect on the challenges of ensuring research rigour in ethnography. Refer to concepts of validity, reliability and representativeness in your answer.

   Extension: Review the studies for their credibility, dependability and transferability.

4. Reflect on the positionality of each of the observers in the studies in this chapter. In each case, how did positionality impact the researcher's ability to gain access to the field and gain the trust of participants? Did any studies raise issues around positionality?

5. What ethical issues do observational studies raise? How have the different researchers in this chapter tried to mitigate ethical issues? Have they done so successfully?

Advantages of observational research:

- Observational research is said to have validity because behaviour is being observed in a natural setting.
- It provides detailed research into a particular microcosm of social life.
- Ethnographic research often supplements observations with interviews to provide detailed accounts of people's lives in a wider context.
- Observations enable us to see people acting in a context, rather than relying on what we are being told about in an interview. In this way, observations may have even higher levels of validity than interviews.
- Sometimes observations can be captured quickly and efficiently using structured observation sheets.
- The representation of the data is often written in a story-like way, making the research compelling for wider audiences and therefore possibly more socially impactful.

Disadvantages of observational research:

- Observations are generally undertaken by one researcher, which compromises reliability.
- Observations are very difficult to repeat.
- They often rely on field notes that are drawn up from memory.
- Ethnographic research is often represented as story-like, which may appear to be less scientific.
- Observational research can be very time-consuming.
- Studies using observations often focus on one small microcosm of society, so they are not representative of the wider population. This does not mean, however, that the studies are not representative of the groups of people being studied.
- Unless researchers use observation sheets that allow them to capture statistical data, it is hard to make inferences of patterns from observations recorded in field notes.

CHAPTER 7: DOCUMENTARY METHODS

What are documentary methods?

A key human activity that separates us from other animals is the ability to make a record of our spoken words and our thoughts, desires, emotions and plans. By writing and drawing, humans have endlessly documented their lives. This means that there are reams of written words, from official government documents, personal diaries and letters, newspaper articles and scientific studies, to organisational newsletters, policies and records. In addition, there are films and their scripts, the recordings of televised content, transcripts of radio recordings, and endless photographic libraries. This all provides a plethora of documentary archives that sociologists can use as sources of data to throw light onto the ways that people articulate and present their experiences of living in the world.

Considering that people spend on average two hours and 27 minutes per day on social media, this is a worthy area for social research, as well as another source of data. Add to this the burgeoning number of blog posts and podcasts, and it is clear that there are many new and accessible archives that can be subjected to documentary analysis. This 'big data' has revolutionised sociology.

Essentially, anything that can be read or assessed for its content can potentially be subjected to sociological analysis. Considering that the ethical dilemmas associated with accessing these types of data are usually minimal, engaging with documentary methods is often a popular choice. Nevertheless, the same ethical principles apply in documentary methods as any other – a researcher must ensure that they have permission to access the documents, that they have the consent of the author of the material, and that the anonymity of the person who produced the document or others that are discussed is preserved.

Thinking further

Sociologists must not only reflect on how new digital platforms may require new research methods, they must also reflect on new and shifting ethical issues that arise as a result. The British Sociological Association has commented on digital ethics as follows:

> Working with digital platforms, networks, and data often raises many new ethical concerns and unanticipated dilemmas. For example, we have to rethink concepts of informed consent and confidentiality (including anonymity), work with new, messy and often confusing definitions of the private and the public and resolve unprecedented tensions between the researcher and the researched. [...] Our position is that our inviolable duty of care to our research participants, and to ourselves, should be reflexively applied using '**situational ethics**' that can allow for discretion, flexibility, and innovation.

 Cleanfluencers

Casey and Littler (2021) examined the rise of the online 'cleanfluencer' Mrs Hinch, an Instagram star with over four million followers. The researchers showed how housework is being reframed as fun, glamorous and therapeutic, and argued that posts like this on social media platforms reinforce patriarchal capitalism and fail to challenge the tedious and exploitative sides of housework. They undertook a discourse analysis of posts and stories on Mrs Hinch's professional Instagram page (@mrshinchhome) between January 2020 and January 2021. They also analysed TV and radio interviews, advertisements and published books.

Key concepts

situational ethics the ethical approach that moral judgments are made in the social context of the entirety of a situation; this differs from an ethical stance that requires the researcher follow a set of prescriptive rules regardless of context

Research tip

You can view the British Sociological Association's website for a discussion on digital ethics: https://www.britsoc.co.uk/about/latest-news/2017/april/ethics-of-digital-research/.

You can also visit the BSA's 'digital annexe' for further information and guidelines on digital ethics: https://www.britsoc.co.uk/media/24309/bsa_statement_of_ethical_practice_annexe.pdf.

Questions

Do you think that social media is valid and reliable data? Give reasons for your answer.

What checks and balances should sociologists use to enhance the rigour of their research in this area?

How do sociologists use documentary methods?

Social researchers can look at contemporary and/or historical sources. Documents can be divided into five main types:

- official (for example, government policy)
- private and personal (for example, diaries, letters, autobiographies)
- mass media (for example, newspapers, magazines)
- visual (for example, photographs, advertisements, art)
- virtual (for example, social media, blogs, websites, chatrooms).

There are two main types of documentary analysis that sociologists use:

- **Content analysis (quantitative):** Content analysis of documentary data tends to focus on counting how many times certain words or phrases are used and subjecting this data to statistical analysis. This is also sometimes called formal content analysis. For example, Mijs and Savage (2019) used Google Ngram to see how many times the terms 'plutocracy' and 'meritocracy' appeared in English language books between 1950 and 2000. Their research showed the disappearance of the former in comparison to wider use of the latter.

- **Discourse analysis (qualitative):** A discourse is a way of thinking, acting or knowing about the world. Discourse analysis looks for themes – it seeks to reveal the hidden meanings behind the words and images. It therefore captures qualitative data.

> **Activity**
>
> Imagine you want to use documentary methods to examine what goes on in schools. Make a list of all the types of documents you might be able to find. Give an example of each type.

🔍 Researcher insights

In 2020, I (Sarah), along with my colleagues Jennie Bristow and Anwesa Chatterjee, conducted a research project about what it is like to study at university and how the relationship between students and academics is changing. We undertook interviews and focus groups, but we also thematically analysed data from the Mass Observation Archive and carried out documentary analyses of seven easily accessed key policy documents. Each team member carefully read and reread these documents, comparing the emerging themes and changing scripts about how the student and the academic were constructed. The team tracked how the discourses of **marketisation**, **consumerism**, individualisation and competition – all features of **neoliberalism** – had gradually come to dominate the policy papers.

The documentary analysis of policy papers showed how the student was increasingly constructed as a consumer and the academic had disappeared from policy discussion, except when cast as a service provider, offering pastoral support.

Analysis of the policy documents was the most straightforward part of the research, requiring no travel and minimal ethical approval. It enabled us to identify dominant public discourses, which could then be checked in interviews and focus groups to see if these ideas were palatable to, and expressed by, students and academics.

Researchers undertaking documentary methods can use multiple types of documents in their analyses, and can do so over a longer time period by including historical social research. In a research project into the criminalisation of fighting sports, I (Jennifer) used a discourse analysis to examine pamphlets, newspaper reports and popular literature, as well as legal cases and statute (law) from the late eighteenth century to the present day. This produced an extensive archive of documents that I analysed in order to look at the way that the sport of boxing had been able to avoid criminalisation, while its older siblings of duelling and prize fighting had not. I examined all these documents in relation to changing social contexts (such as the rise of **medicalisation** and policing) and was able to show that boxing 'sanitised' itself by appearing scientific (rule bound) and 'healthy' (promoting good health and focused on shielding the body using protective equipment). Boxing therefore differentiated itself from duelling and prize fighting. Accessing some historical documents was tricky and it involved long hours of trawling through archives. Fortunately, lots of these archives are now digitised so you can access them from your desk.

Case study: Representations of British Muslims in the news

In Poole and Williamson's 2021 study 'Disrupting or reconfiguring racist narratives about Muslims? The representation of British Muslims during the Covid-19 crisis', the researchers examined British newspaper coverage of Muslims during the first wave of the COVID-19 pandemic. Muslims are usually negatively portrayed in the UK media, and the researchers were interested to see if these depictions changed during the pandemic, considering that many key workers came from ethnic minority communities. Indeed, four of the first doctors to die of COVID-19 in the UK were British Muslim.

Using critical discourse analysis, Poole and Williamson focused on the composition of articles, content, sentence construction and forms of address – the words and phrases used in articles to address audiences as an implied reader. For instance, they suggest a direct address (using the pronoun 'you') is often used to create an 'us and them' mentality. They wanted to explore which dominant narratives prevailed in the newspaper articles to examine whether there was evidence of British Muslims being portrayed as heroes or victims, or as good or bad.

To find their newspaper sample, the researchers used the search terms 'Muslim' and 'Islam' in articles published between 30 March and 30 April 2020. This technique yielded 219 articles, of which 99 (45%) were COVID-19 related. The articles were evenly split between national and international news stories.

Table 1 establishes that the *Daily Mail* was the newspaper to focus most on British Muslims, followed by *The Telegraph*.

	Daily Mail	*The Telegraph*	*The Sun*	*The Mirror*	Total
Local news	35	35	16	19	105
International	45	39	16	14	112
Total	80	74	32	33	219
Covid-related articles	48 (60%)	26 (35%)	15 (46.8%)	10 (30%)	99 (45%)

Table 7.1: Number of articles referencing Muslims and Islam 30/3/2020 to 30/4/2020. Source: Poole, E. and Williamson, M. (2021). 'Disrupting or reconfiguring racist narratives about Muslims? The representation of British Muslims during the Covid crisis'.

Terrorism was the most significant focus in press representations of Islam, followed by articles examining Ramadan. Other key findings from the research were as follows.

- Articles generally failed to identify the ethnicity of a doctor when they died of COVID-19 if they were Muslim. In contrast, Muslim identity was identified in other news stories. The researchers noted: 'The fact that a large number of the victims of Covid-19 were Muslims was largely obscured […] 'Muslim' as an identifier is irrelevant unless it signifies difference.' (p. 276)
- There was a shifting balance towards more negative stories: 'Positive narratives about Muslims supported initially by stories about community initiatives (such as food bank donations and volunteering) […] gave way to macabre and sensationalist imagery of a build-up of coffins at mosques, accompanied by language such as "chilling", "distressing" and "sobering".' (p. 271)
- There was a trend within the newspapers to blame British Muslims for the spread of COVID-19, suggesting the emergence of a new moral panic.

Overall, the authors suggested that the COVID-19 crisis did not see a challenging of dominant discourses about Muslims in the UK. Press coverage continued to reinforce the 'othering' of British Muslims.

> **Questions**
>
> Reflecting on this case study, why were documentary methods important for revealing these findings? Would other methods have been as useful? Why, or why not?

How rigorous are documentary methods?

Before assessing the documentary method itself, it is important to establish the credibility of the source a researcher has used.

Scott (1990) provides four sets of criteria for assessing the status of documents:

- **Authenticity:** Is the *source* genuine (i.e. real)? Who produced the document and why?
- **Credibility:** Is the *data* presented in the source true? Is the material genuine or did the person who produced the document want to make a particular statement about the world? Can you corroborate the content of the document by using other sources?
- **Representativeness:** Is this a typical presentation of data in this type of source?
- **Meaning:** Is it easy to understand what the document means? Can you know that the author of the source intended it to be read or interpreted in a certain way? Are different interpretations of the document possible?

As with any research method, the researcher needs to assess whether the documents are reliable – do they represent the wider field, and can the researcher be sure that other researchers would identify similar themes if they were to analyse the documents?

In terms of validity, a researcher must be confident that the documents are credible and are free from bias and error. When analysing newspaper articles, for example, you would need to take account of whether the paper was politically **left wing** or right wing and whether it was a **broadsheet** or a **tabloid**. Additionally, you would need to be confident that the material was authentic – in documents the author can exaggerate or be focused on establishing their viewpoint.

Key concepts

critical discourse analysis a qualitative approach to analysing documents, which involves describing, analysing and interpreting text and images to explore how discourses construct, maintain and legitimate social inequalities

Foucauldian Discourse Analysis (FDA) a form of discourse analysis that draws on the work of Michel Foucault, who regarded discourses as dominant knowledges circulating in the social world; researchers using an FDA approach focus on revealing how language and visual images used in documents construct or maintain dominant discourses, as well as possibly challenging them

Which sociologists are likely to use documentary methods?

Content analysis tends to be used by positivist social researchers, while discourse analysis tends to be employed by realist, interpretivist and constructionist researchers. **Critical discourse analysis** is underpinned by realist research assumptions. Typically, those undertaking critical discourse analysis seek to reveal hidden meanings and power structures implicit in the texts that they are analysing. Constructionists often employ these discourse analysis methods, too. However, depending on their theoretical perspective, they may focus on dominant discourses in texts – you may sometimes see the term **Foucauldian Discourse Analysis**, which is a mode of interpretation influenced by the work of Michel Foucault.

Thinking further

While 'discourse' is typically used to refer to the analysis of language, Foucault used the term to refer to the relationship between language, knowledge and power. In this respect, a discourse analysis would involve revealing how the words or images presented in documents are circulated as part of wider 'dominant discourses' that govern human conduct and behaviour.

CLASSIC RESEARCH: DISCOURSES OF CRIME

Two contrasting 'classic' studies use documents quite differently to provide sociological insights into understandings of crime.

Study 1

Foucault, M. (1995). *Discipline and Punish: The Birth of the Prison*. New York: Vintage Books.

Study overview

In the opening pages of Michel Foucault's *Discipline and Punish*, we are introduced to a description of the public execution of Damiens, a French domestic servant who had attempted to kill King Louis XV. He was the last person to be executed in France by dismemberment. Foucault quotes the description of the execution from his documentary source, *Pièces originales*:

> On 2 March 1757 Damiens the regicide was condemned 'to make the amende honorable before the main door of the Church of Paris', where he has to be taken and conveyed in a cart, wearing nothing but a shirt, holding a torch of burning wax weighing two pounds'; then, 'in the said cart, to the Place de Greve, where, on a scaffold that will be erected there, the flesh will be torn from his breasts, arms, thighs and calves with red-hot pincers, his right hand, holding the knife with which he committed the said parricide, burnt with sulphur, and, on those places where the flesh will be torn away, poured molten lead, boiling oil, burning resin, wax and sulphur melted together and then his body drawn and quartered by four horses and his limbs and body consumed by fire, reduced to ashes and his ashes thrown to the winds. (p. 3)

Eighty years after the execution of Damiens, Leon Faucher, a liberal French politician, drew up his prison rules of conduct, a document which Foucault also analysed. These included rules such as regimented start times for the prisoners' day, the hours they needed to work, how they ought to dress, and the conduct of morning prayer. In short, rather than being punished for their crimes through death, Faucher's rules showed that prisoners were now punished through regimens that completely eradicated the possibility for any agency and focused on disciplining the mind and soul.

Method and sample

Foucault used these historical documents, among others, to build his documentary archive, which he then subjected to a discourse analysis. From this, he developed a theory.

Findings

Foucault argued that this archival data shows evidence of a shift in the dominant discourses concerning crime and punishment. Despite alleged 'reforms' of punishment – a shift from brutal corporeal punishment mechanisms to more humanistic punishments like imprisonment – Foucault revealed the dehumanising effects of these newer disciplinary mechanisms by which individual prisoners become docile rule-followers.

Using this documentary analysis, he theorised about the way that disciplinary mechanisms of power, premised on constant surveillance, also operate in other large institutions, such as schools and hospitals. He showed how discipline becomes a key feature of modern power, taking over from traditional sovereign (top down) forms of power and authority.

Evaluation

Foucault's work is ground-breaking in the way that it reveals a shift in the way that power operates from a 'top down' model to a more diffuse operation through discourses. As such, his work is both impactful sociologically as well as theoretically. His work has even influenced social research methods themselves by coining the term 'dominant discourse', which is used in discourse analysis methods today.

While clearly impactful, however, his work is not without critique. In particular, his method is often challenged for lacking in transparency. We do not know how he decided which documents to include in his archive, so it is difficult to judge the authenticity, credibility and representativeness of some of his sources, and this brings the reliability of his method into question.

Study 2

Hall, S., Critcher, S., Jefferson, T., Clarke, J. and Roberts, B. (1978). *Policing the Crisis: Mugging, the State and Law & Order*. London: Macmillan Press.

Study overview

Stuart Hall et al. examined British media coverage of what came to be termed the 'mugging crisis' in the 1970s.

Method and sample

The authors analysed American and British newspaper texts from the late 1960s to the mid-1970s to reveal how the term 'mugging' was imported into British media discourse concerning crime. They analysed press coverage in relation to the crime statistics themselves, using a 13-month sample period to analyse *The Guardian* and the *Daily Mirror*. They also had access to a pile of newspaper cuttings.

> **Question**
>
> Why do you think the researchers chose to compare *The Guardian* and the *Daily Mirror* newspapers?

Findings

Prior to August 1972, the term 'mugging' was not used by the British press or police. It was common in the USA, but in the UK similar types of crimes were recorded as robbery or assaults with intent. However, Hall et al. reveal that the term 'mugging' was used 60 times between 1972 and 1973 in British newspapers. Their media analysis revealed that mugging was framed as a social threat, and that it was attached to particular ethnic groups. The discourses circulating in the media led to disproportionate and reactive political responses, which quoted huge rises in muggings despite the concept being new to English discourse. The table below outlines press coverage of 'mugging' between August 1972 and August 1973.

Interestingly, as Hall et al. argued, mugging did not exist as a formal offence in England and there was no evidence of it in the crime statistics. Yet police superintendents, alarmed by a perceived increase in violent crime, particularly among young people, began to ask the Home Office for stronger powers to combat mugging. The newspaper articles incited momentum in the police to change their practices, which then escalated concerns among ordinary people.

Month/Year	The Guardian (1)	Daily Mirror (2)	(1) and (2) combined	Other dailies	Monthly totals
Aug 1972	5	1	6	3	9
Sep 1972	2	5	7	5	12
Oct 1972	7	18	25	19	44
Nov 1972	5	5	10	13	23
Dec 1972	0	2	2	4	6
Jan 1973	4	5	9	4	13
Feb 1973	0	1	1	7	8
Mar 1973	7	9	16	37	53*
Apr 1973	4	4	8	13	21
May 1973	2	0	2	4	6
Jun 1973	0	5	5	0	5
Jul 1973	0	0	0	0	0
Aug 1973	1	1	2	0	2

* Includes thirty-four stories on the Handsworth case.

Table 7.2: The press coverage of 'mugging' (August 1972 to August 1973). Source: Hall, S., Critcher, S., Jefferson, T., Clarke, J. and Roberts, B. (1978). *Policing the Crisis: Mugging, the State and Law & Order.*

Hall et al.'s analysis was realist because they sought to reveal hidden power structures at play (of capitalism and racism) behind media news reports and, for example, the responses of judges when making sentencing decisions. They argued that the rise in reports of mugging constituted a moral panic, which was used to distract people from more deeply rooted social issues such as poverty and unemployment.

Evaluation

This study gives incredible insights into the way that newspapers shape public and political opinion. It takes a realist perspective and, as such, is focused on revealing the hidden structures of racism and classism in news reporting. The researchers also provide empirical evidence of the sociological concept of moral panic.

Although the researchers had a clear sample frame for their newspaper articles, the use of their own newspaper clippings was less rigorous and difficult for other researchers to check, so this aspect of the study is less reliable. By including both a tabloid and broadsheet newspaper in their sample, the researchers limited possible bias in their findings and mitigated any critiques concerning authenticity and credibility. That both newspapers sampled were more politically left-leaning also minimised the possible variations from journalists' political bias.

> **Question**
>
> What conclusions can you draw from this table?

> **Question**
>
> It is worth remembering that Hall et al. had to buy newspapers every day in order to analyse them, and they could not run a keyword search. How do you think the digitalisation of newspaper archives in recent years might impact the popularity of documentary research as a method?

CONTEMPORARY RESEARCH: FAMILY: DADS USING PREGNANCY APPS

Thomas, G., Lupton, D. and Pedersen, S. (2018). '"The appy for a happy pappy": Expectant fatherhood and pregnancy apps'. *Journal of Gender Studies*, 27(7), pp. 759–770.

Study overview

Thomas and colleagues wanted to examine the representation of expectant fathers in pregnancy applications (apps). They argued that support for parenting has moved online, reflected in an enormous amount of parenting apps that give expectant parents a wealth of information. Most of these apps are aimed at mothers, but very little is known about the apps that target expectant fathers.

Method and sample

The researchers examined the ways that pregnancy apps were presented and advertised to potential app users in two online app stores (Google Play and the Apple App Store) in June 2015. They found a total of 665 pregnancy apps on Google Play and 1141 on the App Store. From this sample, they identified 13 apps for men that provided general information and advice about expectant fatherhood, and nine that gave parenting tips for men. These 22 apps formed the basis of their sample.

They conducted a discourse analysis of the app descriptions, examining aspects such as the title and logos of each app, the app's visual content, verbal overviews of the content and purpose, and sometimes information about the app developers.

The analysis was focused on situating the app content in the wider social, cultural and political context in which it was produced. In this regard, the research used a constructionist lens to consider how language used in the apps was reflective of wider power relations.

The researchers asked a series of questions to guide their analysis of the archive (p. 762). These included:

- What features of the app descriptions are employed by the developers to attract prospective users?
- What appeals do the apps make to both female and male users?
- What assumptions about gendered parenting and about pregnancy and reproduction are reproduced to convey meaning and represent the app as a desirable commodity?
- How are expectant fathers portrayed in relation to a pregnant partner?
- What assumptions and values relating to the concept of the 'good' father and supportive partner to a pregnant woman are reproduced in the app descriptions?
- What are the absences in these representations?

The analysis of each app was carried out individually by three researchers, and interpretations were cross checked.

> **Question**
>
> The sample is clearly representative of apps targeting men, but is the sample representative more broadly? Explain your answer.

> **Questions**
>
> Why do you think the research team decided to analyse all apps individually rather than choosing to 'divide and conquer' and take a proportion of the apps each?
>
> What reflections can you make on this methodological choice in terms of the study's reliability?

Findings

The analysis method involved finding themes, and generating and building theories. Overall, the researchers found three key themes:

- **Informing and educating fathers:** They found that apps generally represented fathers as being less committed to parenting than mothers, uncertain of their role and lacking in knowledge about key aspects of new parenthood. In addition, the researchers found gendered assumptions that men needed advice around DIY aspects of parenthood, such as assembling cots.
- **Encouraging men to be attentive and supportive:** The researchers found that a common theme throughout the apps was advice to men around the need to emotionally support their partners in pregnancy and after birth. Pregnant women were framed as being emotional and lacking confidence (particularly body confidence), and men were encouraged to provide emotional support.
- **Ludifying (the mocking of fatherhood):** The authors noted the more prevalent use of humour in men's apps.

The researchers also noted that all three themes were underscored by gendered stereotypes and assumptions about the role of fathers compared with mothers. Expectant fathers were often depicted as inept, foolish and/or reluctant. In sum, the discourse analysis revealed that apps followed 'culturally and historically specific "scripts" that portray fathers as floundering, feckless and flawed' (p. 768).

Evaluation

Drawing on a constructionist approach, the study gives interesting insight into discourses surrounding masculinity and fatherhood, and continuing inequality in the way that parenting is presented. Although the sample is small, it is as representative as it could be, bearing in mind that the researchers looked at all apps pertaining to fatherhood. It is reliable because three researchers undertook the research and cross-checked each other's findings – a form of **investigator triangulation** (see also Chapter 9). Anyone else could look at the archive and see if they agreed with the analysis. As the research only looks at apps, we cannot be certain that it is a valid representation of the way that all contemporary documents present fatherhood. As with all documentary methods, the ethical concerns are minimised.

Key concepts

investigator triangulation
a method in which more than one investigator is used during a research study to cross-check interpretations and minimise bias

🔗 CONTEMPORARY RESEARCH: CRIME: FEMICIDE IN THE NEWS

Taylor, R. (2009). 'Slain and slandered: A content analysis of the portrayal of femicide in crime news'. *Homicide Studies*, 13(1), pp. 21–49.

Study overview

Taylor wanted to examine how femicide victims were portrayed in the media. Femicide is the killing of women – what Taylor describes as the ultimate act of violence against women.

Method and sample

She conducted a content analysis of 292 domestic homicide-related articles, representing 168 separate cases of domestic homicide published by one Florida metropolitan newspaper from 1995 to 2000. She regarded newspaper analysis as socially significant because the media typically portrays violence against women as somehow the fault of the women themselves. Taylor wanted to answer two research questions:

- How were the victim and perpetrator portrayed in the articles? Specifically, what language was used and what behaviours or conditions were disclosed that could lead to a negative or positive portrayal?
- Was a context of domestic violence revealed? If so, how, and what were the sources for this information?

She used a coding sheet to analyse each article. She chose to focus on ten aspects of the article:

1. Title
2. Negative adjectives/behaviours attributed to victim
3. Negative adjectives/behaviours attributed to perpetrator
4. Positive adjectives/behaviours attributed to victim
5. Positive adjectives/behaviours attributed to perpetrator
6. Was any domestic violence context referenced (including no known history of such violence)?
7. Sources referenced for domestic violence context
8. Physical, mental or other pathological issues concerning victim
9. Physical, mental or other pathological issues concerning perpetrator
10. Tone of story

Questions

Why do you think Taylor used a coding sheet? How useful do you think her chosen codes were? Would you have included anything else?

What theoretical perspective do you think that Taylor uses?

Findings

A key theme running throughout Taylor's findings was 'victim blaming' and how the media frequently depicted women who had been killed as victims based on being 'greedy, threatening, sexually promiscuous, […] and rebellious' (p. 34). She showed that the media therefore perpetuated stereotypical images of women victims of crime, gendering them.

Evaluation

Taylor's quantitative content analysis positions her as a positivist researcher. The fact that she was interested in revealing hidden patriarchal assumptions that sat behind media depictions of women shows that she was also realist in her approach. Taylor's sample size was quite large and her analysis used a coding sheet, which arguably increases the validity of her results as it enabled her to be systematic. It also enhances the study's reliability, as the sheet could be used by other researchers.

However, Taylor only used one metropolitan newspaper, printed in one US state (Florida), so the representativeness of her sample can be questioned.

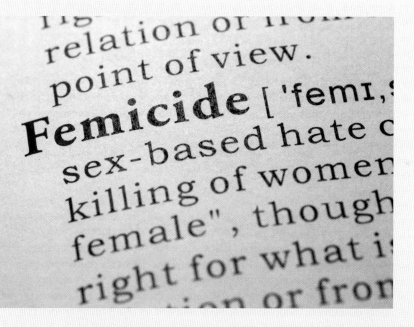

CONTEMPORARY RESEARCH: EDUCATION: EDUCATIONAL POLICY IN ACTION

Ball, S., Maguire, M. and Braun, A. (2012). *How Schools Do Policy: Policy Enactments in Secondary Schools.* London: Routledge.

Study overview

Ball et al. wanted to examine how educational policy was interpreted and dealt with in schools.

Method and sample

They conducted four case studies in English state secondary schools over two years. While they used multiple methods, they conducted an extensive discourse analysis of what they termed 'hard' educational policy texts as well as other visual data – including photographs and visual displays, which they termed 'artefacts' – that they found within the school. The researchers then subjected these documents to a discourse analysis method, drawing inspiration from Foucault.

They described educational policies as 'discursive formations' – that is, 'sets of texts, events and practices that speak to wider social processes of schooling such as the production of "the student", the "purpose of schooling" and the construction of "the teacher"' (p. 123).

The authors included visual artefacts in their analysis, which they noted are often missing from documentary methods and policy analysis more specifically.

Findings

The researchers revealed how the secondary school interprets educational policy to reinforce the dominant idea of the 'good student' in various ways, including through visual signs. They argued that a 'student of the week' display is one example of many that demonstrates the 'good student' as the one who displays virtues of discipline, commitment and enthusiasm. Other examples are notice boards that outline to students how to earn top grades, which work to normalise policy discourses of educational achievement and success through grades, but silence and omit the 'failing' students from the educational institution.

> **Question**
>
> What do you think including the visual materials adds to the picture that Ball et al. paint of the way that educational policy is embedded in school life?

In another example, the researchers note the way that some department rooms and staff rooms in one school displayed lists of targeted underachieving students with their photographs alongside. Staff were encouraged to stick post-it notes on these photographs noting any positive progressions of students, urging them to 'keep in mind' the students and to encourage one another to attend to them.

All this demonstrates how policy is visualised in a school environment and how documentary methods such as those employed here enable the researchers to focus on analysing the messaging behind policy enactments. They reveal deep relations between policy documents, educational settings and discursive productions of power. These are used to ensure that students and teachers govern their own conduct in line with broader norms of education – those of success and hard work, all underpinned by the 'good student' and 'good teacher' discourses.

Evaluation

The research tries to address a gap in knowledge of how educational policy is enacted in schools, and uses innovative methods of documentary analysis to do this. However, the validity and reliability of the study is difficult to discern because there is relatively little evidence of how the documentary analysis itself was carried out. The researchers assert their findings without a clear outline of how they were achieved. This is a common issue in many documentary methods.

CONTEMPORARY RESEARCH: INEQUALITY: POVERTY IN THE NEWS

McArthur, D. and Reeves, A. (2019). 'The Rhetoric of Recessions: How British Newspapers Talk about the Poor When Unemployment Rises, 1896–2000'. *Sociology*, 56(5), pp. 1005–1025.

Study overview

The researchers' objective here was to measure changes in the amount and form of stigmatising language used about people living in poverty during the twentieth century.

Method and sample

In this study, the researchers examined the *Mail* (*Sunday* and *Daily*), *the Telegraph*, *The Times*, the *Financial Times* and *The Economist*, but recognised that these are all right-wing newspapers. They noted the frequency of descriptors of poor people and assessed which words were used at different points in time.

Findings

You may have noticed the declining popularity of the use of the term 'pauper', and the increasing popularity of the terms 'underclass' and 'scrounger'. The table also suggests that stigmatising words are more likely to be used when unemployment prevails. The researchers suggest, therefore, that the media has a strong role in shaping public attitudes towards people in poverty, and these right-wing newspapers perpetuate negative views of the poor.

Evaluation

The authors only analysed right-wing newspapers, which raises issues of representativeness and also the validity of claims that British newspapers more broadly (as the sensationalist title suggests) are stigmatising about poor and unemployed people in the UK. The research study is therefore potentially ideologically motivated rather than balanced and unbiased.

Word	Average	Min	Max	Sparkline
Peasant*	948.30	467	2413	
Tramp*	659.21	195	1216	
Beggar*	332.31	46	1313	
Peon*	305.90	46	587	
Pauper*	175.58	10	844	
Dependency	103.70	14	511	
Idler*	85.61	22	202	
Lower class	72.03	8	181	
Delinquent	62.16	7	272	
Loafer*	56.02	4	201	
Delinquency	55.58	2	242	
Indigent	50.02	2	241	
Vagrant	40.94	4	138	
Feckless	39.27	0	216	
Indolent	30.60	5	85	
Unemployable	24.62	0	80	
Vagrancy	23.47	0	93	
Underclass	22.19	0	225	
Deserving poor	22.00	0	153	
Shirker*	19.19	1	230	
Scrounger*	16.80	0	130	
Residuum	10.72	0	52	
Skiver*	9.71	0	67	
Dependent on benefits	7.86	0	45	
Criminal class	7.30	0	19	
Workshy	6.64	0	39	
Dangerous class	5.20	0	15	
Riff raff	1.66	0	20	

Notes: Data from Gale Newsvault. Words ordered by descending average frequency. Sparklines are intended to give a sense of relative frequency of word across time (1896–2000) and should be read with respect to minimum and maximum for each word.

Table 7.3: Descriptive statistics for words measuring stigmatising rhetoric about the poor. Source: McArthur, D. and Reeves, A. (2019). 'The Rhetoric of Recessions: How British Newspapers Talk about the Poor When Unemployment Rises, 1896–2000'. *Sociology*, 56(5), p.1011.

Question

In this table, you can see not just descriptive statistics for each term, but also sparklines to show how often these words are used in the period 1896–2000. Which words are more popular in the early twentieth century, and which are more popular in the late twentieth century?

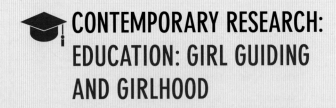

CONTEMPORARY RESEARCH: EDUCATION: GIRL GUIDING AND GIRLHOOD

Halls, A., Uprichard, E. and Jackson, C. (2018). 'Changing Girlhoods – Changing Girl Guiding'. *The Sociological Review*, 66(1), pp. 257–272.

Study overview

The Girl Guides began as a grassroots response to the Boy Scouts, which had been founded by Lord Robert Baden-Powell in 1907. The Scouts was designed to train boys in 'appropriate citizenship', as future leaders of the British Empire. While the Scouts appealed to girls – and girls had tried to join or align themselves with Scouts – there was some resistance based on whether this demonstrated appropriate **femininity** versus **unacceptable femininity**. The Girl Scout movement emerged in parallel to the Boy Scouts, in line with feminine social ideals. Baden-Powell's sister, Agnes, who shared her brother's conservative ideals, headed this movement, changing the Girl Scouts' title to Guides to reflect more feminine conventions. Halls et al. wanted to know how girlhood has changed since then.

Method and sample

The researchers looked at all editions of Girl Guide handbooks between 1910 and 2011. They accessed these through the National Archives in London. Mostly (although not always) they were annual publications; the researchers examined 87 handbooks in total. They used qualitative thematic analysis, which involved colour coding text by hand. They focused on recurring or different themes or images, looking at the 'form and content' of the handbook, and considering the social context in which the texts were produced.

Findings

They found three main strands of change in the handbooks:

- Girls were seen to be increasingly competent, but were much more likely to be constructed as vulnerable and 'at risk' – the **risk discourse** was more prevalent in recent handbooks, as was the need for adult supervision.
- The notion that girls needed their own space for guiding and that they were 'not boys' was emphasised throughout all handbooks. Girlhood was constructed as what it is 'not to be a boy', emphasised through different feminised activities, such as cooking and baking. In the handbooks, girls were portrayed as 'becoming women' – as future wives and mothers.
- Enjoyment was a key theme throughout all the handbooks and did not appear to change over time.

Overall, the researchers found that girlhood and childhood are socially constructed categories. However, there was some consistency in how girls are expected to behave, what they can do and what they may become.

Evaluation

The research showed the enduring stereotypes of femininity that continue to be reflected through Girl Guiding today. The study can be described as representative, as the researchers looked at all editions of handbooks that were available. Because the documents are in the public domain, the results can be corroborated by anyone, so can be considered reliable. There were few ethical issues with this study because documentary analysis does not involve human subjects. However, the researchers did note that one challenge of using historical documentation like this is that the Girl Guides, and the writers of the handbooks, could not speak for themselves and confirm or deny the interpretations made by the researchers.

Documentary methods as a social research method

1. Look at the range of documents that social researchers have used in this chapter. Which do you think are the most valid, reliable and representative? Why?
2. How does a social researcher's theoretical orientation influence their approach to documentary methods? Illustrate your answer with examples from this chapter.
3. Why do you think documentary methods are a popular choice for sociologists?

Advantages of documentary methods:

- They are relatively easy to access; indeed they are sometimes described as 'armchair' sociology, because research can be undertaken from home, without going into the field.
- They are inexpensive to conduct.
- There are minimal ethical issues associated with them.
- They are reliable, because many documents are in the public domain and therefore can be accessed and analysed by multiple researchers.
- Documentary methods often enable longitudinal research – for example, to see how representations change over time.

Disadvantages of documentary methods:

- Despite minimal ethical issues, documentary methods do raise some ethical concerns, particularly around informed consent, such as when analysing private documents or anonymous online archives.
- Despite being easily accessible, documentary methods are less reliable because texts are open to multiple interpretations.
- Information may be biased or inaccurate – for example, newspaper sources are often politicised, so an analysis cannot necessarily claim validity.
- Data may not be representative – for example, an archive of personal blogs may not contain the full range of views on a topic. Making generalisations is therefore more difficult.

CHAPTER 8: EXPERIMENTS

What are experimental methods?

Imagine you want to find out whether teachers treat children differently in class, whether employers are biased towards certain ethnicities when recruiting for jobs, or whether people break the law more often when they think they are not being watched. Some sociologists use 'experimental designs' to test theories and answer questions such as those posed above. Sometimes sociologists will even set up social situations to manipulate events/factors (variables) and then observe what happens.

> ### Case study: Gender bias experiment
>
> In 2019, a group of researchers sent out fictitious curriculum vitae (CVs) to employers to see whether they were biased against hiring women. They found that when the candidates were of equal merit, men were favoured. Although the bias was reduced when women were more qualified for the job, it increased if they had children (Gonzalez et al., 2019).

The term 'Hawthorne Effect', which is used to describe situations where participants in a study change their behaviour based on what they think the researcher wants, derives from early sociological experiments. These experiments, held in Western Electric's Hawthorne Works in the 1920s and 1930s, deliberately manipulated different working conditions and examined whether changes such as different lighting and work breaks in a factory affected the workers' productivity. The study revealed that it was not the manipulation of the variables that impacted productivity, but the presence of the researchers themselves. Workers were more productive when they were being watched.

The 1930s were the heyday for the use of experiments in social sciences and sociology, particularly in the USA. However, due to several controversial and ethically suspect studies, their popularity has since dwindled. Sociologists are also critical of experimental research for another reason, one that is central to the disciplinary focus of sociology itself: sociology aims to study human actions in natural settings, but experimental designs rely on some form of control being placed on human action – whether in a laboratory or in a natural setting (in the field). Therefore, any research on actions observed in a controlled setting will not necessarily replicate 'real social life' and may produce unnatural and therefore invalid responses.

Case study: The ethics of human experiments

A number of controversial experiments have made sociologists cautious about using this method.

The Tuskegee Experiment (1932–72)

Beginning in the 1930s, the United States Public Health Service and the Centers for Disease Control and Prevention carried out a longitudinal experiment on African Americans with syphilis, based in Tuskegee. Researchers in this epidemiological study wanted to find out about the progress and impact of the disease. Six hundred African American men – 399 with syphilis and 201 without the disease – were enticed into the study, without informed consent, via the offer of free medical care and food. Penicillin became widely available in 1943 and was successfully used to treat the disease; however, the men in the study were not offered this antibiotic because the researchers wanted to learn more about the course of the disease if syphilis went untreated. The study gained widespread public attention and criticism when this withholding of treatment was revealed.

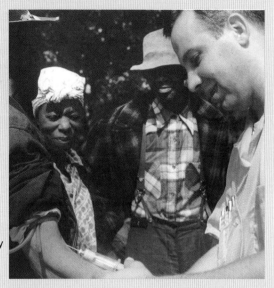

Although this is a case study within the field of health and epidemiology rather than sociology, it exposes the ethical dilemmas of experiments.

Stanford Prison Experiments

In his famous 1971 social psychology Stanford Prison Experiments, Zimbardo (1973) recruited university students to a study designed to examine obedience and authority. He wanted to understand how the atrocities that took place during the Second World War were perpetrated by human beings. In particular, he wanted to see whether those in positions of authority who carried out brutalities were evil, or whether environmental factors may have impacted their behaviour. Would everyone behave in the same way if put in the same situation?

The study recruited 24 male volunteers, who were paid a daily rate for their participation. In a simulated prison environment, participants were randomly selected to be either a prisoner or a guard. Zimbardo found that each individual took on the role they were allocated, with guards using increasingly brutal methods to keep their mock prisoners in check. The study showed that when people were put into positions of authority, they would devise and impose humiliating acts of punishment and violence.

The harms of this study – to the mental and physical wellbeing of the mock prisoners, but also to those who were acting as guards – are obvious. However, Zimbardo justified his research, claiming that it revealed important insights into the ways that people readily embrace and perform the social roles expected of them, even when they break moral codes.

Questions

Do you agree with Zimbardo's justification for the experiment?

Could these insights about the effects of perceived power on individuals have been gained in any other way?

Do you think the harm outweighs the benefits, or vice versa?

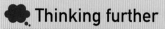 Thinking further

There are several ethical dilemmas involved in social experiments:

- In many experiments, researchers make a distinction between a control group and an experimental group. They often withhold treatments or interventions from the control group, and therefore this is inequitable.
- A lack of informed consent – experiments are premised on the participants not knowing which group they are in, which compromises their ability to give consent.
- Historical experiments that manipulate human action have had harmful effects on participants, although these effects were unintentional.

Gender and housework norms

Thebauld et al. (2021) wanted to study the gendered expectations of housework. They recruited 646 respondents through Amazon.com (where customers can sign up to take part in paid research studies). Researchers showed participants photographs of messy rooms and told them the gender of the occupier. They were asked how messy they thought the room was and how urgent it was to clean it up. The researchers tested two hypotheses:

- Female observers would perceive rooms to be messier and in more urgent need of cleaning.
- If the room was deemed to be a female room, the occupants would face more negative social reactions than if they were male.

The results showed that the first hypothesis was wrong – men and women perceived messiness in the same way. The second hypothesis was upheld – women were judged more harshly for having a messy room, as it was generally felt that women should be responsible for keeping their rooms clean.

How do sociologists use experimental methods?

Social researchers use three main types of experimental method:

- **Laboratory:** This is the least likely method to be used, due to issues of validity (see below). It involves deliberately manipulating variables in a controlled setting. Because the setting is artificial, and because social researchers are interested in understanding naturally occurring social phenomena, laboratory experiments are less popular.
- **Field:** Field experiments take place in real-world or natural settings. However, researchers still try to manipulate behaviour and observe changes based on their interventions. These experiments are more imprecise than those in laboratory settings, because the complex nature of social life cannot be controlled or predicted.

- **Survey:** This is a more common type of experimental design used in sociology. It involves using statistical data to test hypotheses. Researchers can produce an experimental design in surveys to measure different attitudes to social phenomena (see Housework study above).

> ### Case study: Rosenthal and Jacobsen's self-fulfilling prophecy and education
>
> A famous example of a field experiment was conducted by Rosenthal and Jacobsen (1968). They wanted to test whether teachers' expectations of students impacted educational performance. The researchers gave the teachers a list of students who were about to experience an intellectual growth spurt on the basis of test results. However, this test was non-existent, and the list of students' names was picked at random. The study revealed increasing IQ scores in the students who were listed. The researchers concluded that there was a powerful impact of the self-fulfilling prophecy – when teachers had positive expectations of students, they performed better educationally. However, the study was criticised because teachers, students and parents were not told about the research, so they could not give informed consent.

How rigorous are experimental methods?

Laboratory experiments are said to have high levels of control and therefore internal validity. However, many sociologists argue that it is impossible to replicate natural social settings. Therefore, experiments have lower levels of **ecological validity** (that is, they do not necessarily apply to the real world), which lowers external validity overall (they are not generalisable). Laboratory experiments are often said to be reliable because they are repeatable.

Field experiments may be regarded as more ecologically valid, but researchers have less control over external variables (lower internal validity) because the research takes place in a natural rather than in a laboratory setting. Field experiments are not always reliable, as they are harder to repeat.

Survey experiments are as reliable and valid as the data and survey instruments being used. See Chapters 3 and 4 for more detail on official statistics and surveys.

Which sociologists are likely to use experimental methods?

Positivist sociologists are more likely to use experiments, as they aim to replicate natural scientific methods. Sociologists using experimental methods look at cause and effect, or correlational relationships between different variables. Overall, interpretivist, realist and constructionist sociologists are less likely to use experimental methods. However, interpretivist sociologist Garfinkel is an outlier (see Classic research below).

CLASSIC RESEARCH: BREACHING EXPERIMENTS

Garfinkel, H. (1967/1991). *Studies in Ethnomethodology*. London: Wiley.

Study overview

As an interpretivist sociologist, Garfinkel was interested in the ways that human beings artfully accomplish social life, build social order and create shared meanings. He described the ways that people are constantly searching for patterns to make sense of what is going on around them.

Method and sample

He used the term 'ethnomethodology' to describe his approach, which means studying the practices human beings use in the process of living together (people's methods). He argued that to uncover the strategies and methods used to maintain social order, the sociologist had to deliberately unsettle social life and observe what happened next.

Somewhat confusingly, Garfinkel used the term 'documentary method' differently to how we use it in this book. He regarded documentary methods as the techniques that ordinary people use to enable them to cooperate with others. Garfinkel showed how people always try to account for their choices and decisions in everyday life. The documentary method, for Garfinkel, referred to people's everyday practices of grouping things together, making categories of things and actions, based on what they already knew. Human beings actively group together past, present and future events to produce a coherent interpretation of what is going on. In his research, he was keen to uncover how this documentary approach worked to maintain social order.

In one of his studies, Garfinkel asked his students to deliberately unsettle conversational rules when they bumped into their friends by asking them odd questions or by seeking clarification when it was not needed. He referred to this as a **breaching experiment**, designed to demonstrate the unacknowledged presence of social norms and what happens when they are broken. He showed that people try extremely hard to normalise odd behaviour so that social order is restored.

Key concepts

breaching experiment an experiment that intentionally disrupts (breaches) social norms

> **Questions**
>
> What do you notice about the exchange between these two friends?
>
> How would your interpretation of this exchange alter if you thought it was a conversation between a doctor and a patient?

Here is an excerpt from Garfinkel's study.

> S: Hi Ray. How is your girlfriend feeling?
> E: What do you mean, "How is she feeling?" Do you mean physical or mental?
> S: I mean how is she feeling? What's the matter with you? (He looked peeved)
> E: Nothing. Just explain a little clearer what do you mean?
> S: Skip it. How are your Med School applications coming?
> E: What do you mean, "How are they?"
> S: You know what I mean!
> E: I really don't.
> S: What's the matter with you? Are you sick? (p. 42)

Findings

Garfinkel drew attention to understanding conversations and exchanges in terms of the context in which they occur, which he referred to as **indexicality**. Overall, Garfinkel's experiments revealed that social order is fragile and only accomplished by human beings indexically, reflexively and meaningfully. He showed that people work hard to restore social order when it seems to be breaking down.

Evaluation

Garfinkel's experiments took place in natural settings. Though somewhat artificial –they were staged, and required some performances on the part of the experimenters – they were more valid than laboratory experiments. The studies were important to sociologists for understanding the social norms that sustain human interactions. Although the studies required that experimenters deliberately unsettled social norms and conventions, the experiments themselves were not particularly sensitive, although some respondents did suffer some levels of harm (for example, some became agitated). Therefore, the experiments have significant ethical considerations.

 Becoming a social researcher

Garfinkel asked his students to pretend to be a lodger in their own home. He suggested that they might act in ways that were especially polite, cautious, impersonal and formal. As you might imagine, family members were confused and sometimes angry at this behaviour.

These responses indicate that we have a set of assumptions about how family members are supposed to behave. The students in Garfinkel's experiment found that very quickly their parents looked for explanations for the odd behaviour (were the students ill or rebelling?) and either told them off or became sympathetic. Both responses restored social order.

Perhaps you would like to (carefully) try this at home!

CLASSIC RESEARCH: JANE ELLIOTT'S BLUE EYES, BROWN EYES EXPERIMENT

There is no official written record of this experiment, but the following book may be useful: Bloom, S.G. (2021). *Blue Eyes, Brown Eyes*. California: University of California Press. The experiment is also available as a video documentary online.

Study overview

In her televised study of 1968, Elliott was interested to learn how racism operated in schools. She designed an experiment to artificially assess how discrimination occurred and was experienced by school children. She divided her class into two groups: blue-eyed students and brown-eyed students.

Method and sample

She told the class that the brown-eyed students were more intelligent and genetically superior. She found that, astonishingly quickly, the children with brown eyes became more confident in class, sought out other brown-eyed children to play with, and were rude to and excluded blue-eyed children. In contrast, the blue-eyed children visibly became more withdrawn and emotional.

Findings

The social experiment reveals the way that discrimination is built in practice and is socially reproduced. It gives real insight into how discrimination is structured, reproduced and experienced in education.

Evaluation

Elliott's research is socially significant, as it continues to be referred to today as a canonical study that established the operation and social reproduction of racial discrimination. The sample was based on only one group of children in one school setting and is therefore not representative, although we can suggest, based on knowledge of historical and continued racial discrimination, that the findings are valid. Arguably, however, the study would not be repeatable due to the possible harm that research respondents, in this case children, could experience.

CONTEMPORARY RESEARCH: EDUCATION: ETHNIC AND GENDER BIAS IN ONLINE CLASSES

Baker, R., Dee, T., Evans, B. and John, J. (2022). 'Bias in online classes: Evidence from a field experiment'. *Economics of Education Review*, 88, 102259.

Study overview

Baker et al. conducted a field experiment to look at bias in online classes using massive open online courses (MOOCS). The MOOCS were aimed at those of higher education level, although they were open to anyone. Most of the courses were offered by not-for-profit higher education agencies in the USA. The researchers wanted to see whether instructors and peers responded differently to typed comments in the MOOC chat feature based on students' ethnicity.

Method and sample

To conduct this study, the authors used 124 MOOCS covering a range of subject areas, including accounting, computer programming and epidemiology. They created fictional student identities and placed eight discussion forum comments in each of the 124 MOOCS. Eight student accounts were used to place a comment each, and each account was given a name connotative of a specific gender and ethnic origin (White, Black, Indian, Chinese).

The researchers randomly assigned each student comment to the discussion forums and spaced them out evenly throughout the duration of the course, from the beginning to two weeks before the end. They analysed all replies to each comment for two weeks after placement and observed responses by both instructors and peers, looking for differences in the number of responses given to the fictional student groups to see if there were correlations with gender and ethnicity.

> **Question**
>
> What do the findings reveal about the probable relationship between instructor and peer bias and ethnicity and gender in online educational settings?

Findings

These two charts record some of the study's key findings (p. 5). Figure 8.1 shows instructor responses to the comments and Figure 8.2 quantifies peer responses to the comments.

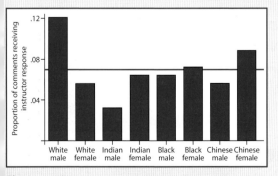

Figure 8.1: Instructor response.

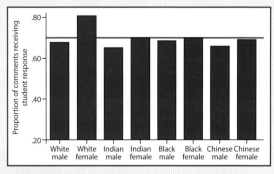

Figure 8.2: Peer response.

As the charts show, the researchers found that White male students received more responses from the instructors and White female students received more responses from peers. Indian males fared worst in terms of both instructor responses and peer responses. Females of all ethnicities fared better than males of all ethnicities in peer responses. From instructors, after White males, Chinese and Black females fared better than White females.

Evaluation

The study importantly expanded knowledge in the areas of digitally mediated online learning, and ethnicity and education. This type of field experiment is useful for sociologists, because it clearly reveals some positive bias from instructors towards White males and from peers towards White females. This supports data from other studies, which highlights the continued presence of ethnic discrimination and racism in education.

The sample was relatively easy to access, as it was an online open module. Because it was a field experiment taking place in a naturalistic rather than artificial laboratory setting, arguably the results are valid. However, a disadvantage of this field experiment is that it is hard to control for all variables, so it is difficult to assert with confidence a correlation between ethnicity and instructor/peer response. Although the researchers tried to control variables by being consistent with when posts were placed in MOOCS (e.g. time of day, spread out throughout the course), other factors may influence responses beyond the researchers' control. The study was reliable as it could be repeated.

Case study: Bias in recruitment

Baker et al.'s study shows that experimental research reveals bias in educational settings. This bias is replicated in field research into bias in employment. Wood et al. (2009) carried out research into hiring discrimination commissioned by the Department of Work and Pensions (DWP). They sent fictitious CVs with different ethnic-sounding names in response to different job advertisements. The research revealed that ethnic minority applicants were less likely to receive a positive response than White applicants.

CONTEMPORARY RESEARCH: FAMILY: GENDER NON-CONFORMITY

Stacey, L. (2021). '(Bio)logics of the Family: Gender, Biological Relatedness, and Attitudes Toward Children's Gender Nonconformity in a Vignette Experiment'. *Sociological Forum*, 37(1), pp. 222–245.

Study overview

Stacey wanted to develop an experiment to reveal normative beliefs held by family members about gender. However, he was conscious that many people would try to give socially desirable responses (**social desirability thesis**) rather than true and valid accounts. As such, he designed his survey using hypothetical vignettes (different made-up scenarios), describing different types of gender non-conformity. He thought this would give a more valid account of people's real beliefs about the gender non/conformity of family members.

He asked two key research questions:

- To what extent do parents' beliefs vary about a hypothetical biological child's or stepchild's gender non-conformity?
- To what extent do these beliefs vary depending on the gender of the hypothetical child?

He hypothesised that parents would be more upset and less supportive when presented with a son rather than a daughter who acted in a gender non-conforming way. He also hypothesised that parents would assume more responsibility, express greater concern over peer-reaction and reveal heightened concerns about judgments from other parents over a non-conforming biological son.

Method and sample

Stacey's survey was open to all US citizens over 18 years old. Participants in the vignette survey were limited to parents.

He advertised the study through MTurk – an online workplace platform through Amazon. Respondents were paid one dollar per survey, which took on average seven minutes to complete. He asked demographic questions first about parenting to secure his sample. A total of 815 people completed the survey, but only 712 of these surveys were usable and could be analysed.

Respondents were randomly assigned to one of four vignette conditions:
- Boy (John) dressed as a princess (1. biological son; 2. stepson)
- Girl (Jennifer) dressed as a firefighter (3. biological daughter; 4. stepdaughter)

> **Key concepts**
> **social desirability thesis** a response bias in surveys, when participants respond in the way they deem socially acceptable rather than providing their own beliefs and experiences; socially desirable responses affect a study's validity

Below is an example vignette used in the study.

> *John, your ten-year-old biological son comes home from school one day and tells you that he wants to be a Princess for Halloween. You are hosting a party to celebrate Halloween and you've invited your friends, neighbors, and their children. John plans to wear the Princess costume to the party and when you take him trick-or-treating the following night. (p. 230)*

After reading the vignette, survey respondents were asked a series of questions about the hypothetical child's request to wear gender non-conforming clothing. Using a Likert scale, they were asked whether they strongly disagreed, disagreed, were neutral, agreed or strongly agreed with the following statements (p. 232):

- *I would be upset*
- *I would be supportive of [John's/Jennifer's] decision*
- *I would feel as if I had done something wrong as a parent*
- *I would worry that other kids might bully or judge [John/Jennifer] harshly for wearing the costume*
- *I would worry that other parents would judge me negatively if [John/Jennifer] were to wear the costume in public or to the party.*

There was also an open-ended narrative response section, where participants could provide additional qualitative comments on the vignettes.

Findings

Stacey's research confirmed both of his hypotheses – parents were more upset if a hypothetical son demonstrated gender non-conforming behaviour, and if this child was biological.

Evaluation

Stacey notes that his survey was not fully representative. He collected demographic information about the respondents, and they were shown to be more educated, less religious and less conservative than the United States population as a whole, and therefore possibly more understanding of gender non-conformity. Another limitation Stacey notes is that because the study was an experiment, there is no certainty that this is how the respondents would behave in real-life situations. Although there is evidence to suggest that there is congruence between these hypothetical scenarios and what people actually think and do, it is important to remember the limitations of experimental research when studying the complex social world around us.

CONTEMPORARY RESEARCH: FAMILY: PERCEPTIONS OF 'ATYPICAL' PARENTS

Quadlin, N., Jeon, N., Doan, L. and Powell, B. (2022). 'Untangling perceptions of atypical parents'. *Journal of Marriage and Family*, 84(4), pp. 1175–1195.

Study overview

This study focused on public perceptions of 'atypical parents' in the USA. The researchers were interested in exploring the normative beliefs and attitudes of the US population concerning parenting, and whether there existed discriminatory beliefs and attitudes towards families who do not meet normative expectations of the 'traditional' or 'typical' family, comprising two biological parents and their offspring. The researchers described families who fall outside this grouping as 'atypical', but they recognised the limitations of this terminology, particularly given the decline of the 'typical family' and the rise of more diverse family types.

Five types of 'atypical' parents were identified:
- single mothers
- single fathers
- lesbian couples
- gay couples
- adoptive parents.

The researchers wanted to know whether, and why, different normative expectations of family structure persist and how this impacted attitudes towards different types of parenting.

Method and sample

The researchers conducted a national telephone survey of 827 adults in the USA. However, not all respondents were asked the same questions; specifically, the researchers divided respondents into five groups, with each group focused on answering questions about their perceptions of parenting of one of the five 'atypical' parent types. Respondents were asked to assess:
- how well the atypical parent type could raise a child
- how well the atypical parent type could provide for a child's basic needs
- how possible it was for the atypical parent type to have warm relationships with their children
- how well the atypical parent type could teach the children about important values.

The phone survey was conducted by the researchers themselves, as well as a trained team of graduate students from the same university. Survey questions were pre-tested and revised based on a pilot study.

The sample was generated by using a list-assisted random digit dialling of landline and mobile phones. Within each household, one respondent was randomly selected. Mobile phones were only contacted if the owner was aged over 18.

The researchers used a series of vignettes to ascertain people's feelings, attitudes and beliefs towards different types of atypical parents. Here is an example of one vignette and the questions that were asked of the category 'single mother':

> People these days have differing opinions as to how well certain groups of people can raise children. I will read some statements about how well certain groups of people can raise children. Please tell me whether you strongly agree, somewhat agree, somewhat disagree or strongly disagree with the following statements.
>
> (1) A single mother can bring up a child as well as two parents together.
>
> (2) A single mother can take care of her child's basic needs, like providing food, shelter, and protection as well as two parents together.
>
> (3) What about having as warm a relationship with her child as two parents together? (Do you strongly agree, somewhat agree, somewhat disagree, or strongly disagree that a single mother can have as warm a relationship with her child as two parents together?)
>
> (4) What about teaching her child important values? (Do you strongly agree, somewhat agree, somewhat disagree, or strongly disagree that a single mother can teach her child important values as well as two parents together?) (p. 1180)

One methodological decision made by the researchers was to assign survey respondents only one category of atypical parent. They gave the following rationale:

> In our case, a respondent might have been asked about single fathers first, and then they might later try to provide equivalent answers about single mothers to maintain internal consistency or avoid the appearance of social desirability bias. The between-subjects design ensures that each respondent gave their initial impressions about only one parent type.

The researchers chose three variables to assess:

- perceptions of economic dimensions of parenting
- perceptions of emotional dimensions of parenting
- perceptions of moral dimensions of parenting.

They also asked people to assess their view overall of how well parents could raise a child.

Questions

What do you notice about these questions? Why does the researcher ask more than one question?

What type of scale do the researchers use to record their responses?

In what other ways could this data have been captured using a social survey?

Why do you think the researchers chose to use vignettes as part of their surveys?

Questions

Do you think this decision makes the study more valid? Why, or why not?

Do you think there are any problems with this rationale? Explain your response.

Findings

Table 8.1 depicts the descriptive statistics generated through the survey research.

	Single mother	Single father	Lesbian couple	Gay couple	Adoptive parents
Can bring up a child					
Mean	3.05	2.92	2.77	2.63	3.70
Strongly agree	0.31	0.24	0.34	0.29	0.77
Agree	0.50	0.49	0.31	0.31	0.18
Disagree	0.14	0.20	0.15	0.14	0.02
Strongly disagree	0.06	0.06	0.21	0.26	0.02
$n =$	169	164	163	160	164
Can take care of basic needs					
Mean	3.20	3.46	3.29	3.20	3.85
Strongly agree	0.43	0.56	0.55	0.55	0.87
Agree	0.40	0.36	0.28	0.25	0.12
Disagree	0.12	0.07	0.06	0.06	0.01
Strongly disagree	0.05	0.01	0.10	0.14	0.01
$n =$	169	165	164	161	165
Can have warm relationship					
Mean	3.58	3.22	3.16	2.95	3.75
Strongly agree	0.71	0.47	0.55	0.54	0.79
Agree	0.20	0.36	0.22	0.24	0.18
Disagree	0.06	0.10	0.07	0.14	0.01
Strongly disagree	0.03	0.07	0.16	0.18	0.02
$n =$	170	163	165	161	165

Can teach important values					
Mean	3.63	3.49	3.12	3.01	3.89
(SD)	(0.68)	(0.82)	(1.12)	(1.20)	(.37)
Strongly agree	0.72	0.65	0.53	0.51	0.90
Agree	0.22	0.24	0.20	0.21	0.10
Disagree	0.04	0.06	0.12	0.06	0.00
Strongly disagree	0.02	0.05	0.15	0.22	0.01
$n =$	170	165	165	161	166

Note: Proportions may not add to 1 due to rounding. Source: Constructing the Family and Higher Education Survey, 2015.

Table 8.1: Descriptive statistics – outcome variables. Source: Quadlin, N., Jeon, N., Doan, L. and Powell, B. (2022). 'Untangling perceptions of atypical parents'.

Question

What points can you extract from this table?

You might have noted the following things that are evident from the data:

- The most accepted 'normal' atypical family type are adoptive parents. Across all questions, they are most likely to be allocated 'strongly agree' responses.
- The next two categories that are fairly evenly distributed are single mothers and single fathers. While neither scores as many 'strongly agree' responses as adoptive parents, they are regarded as better parents across the board than same-sex parents (lesbian and gay).

The research also reveals that:

- women and respondents who have gay friends or relatives are more receptive to atypical parents
- those who are less receptive to atypical parents are often religious
- ethnicity, age and education are all variables that shape attitudes towards atypical families.

Further interesting outcomes include the following:

- When focused on the economic dimensions of parenting, respondents tend to favour adoptive parents as well as single fathers over single mothers and same-sex parents.
- When focused on the emotional aspects of parenting, single fathers and same-sex couples are seen as lagging behind single mothers and adoptive parents. The authors note that the sentiments towards lesbian mothers in particular are interesting given the gender dimension, where respondents clearly favour single mothers over fathers, but not two women raising children.

- When focused on the moral dimension of parenting, same-sex couples again score lower than single mothers and single fathers, with adoptive parents scoring the highest in public perception.
- The main parenting concern for single mothers appears to be their economic ability to raise a child.
- The main parenting concern for single fathers is their ability to have warm relationships with their children.
- Lesbian couples are stigmatised on economic and moral grounds by the public.
- Gay parents are stigmatised in all dimensions – economic, emotional and moral.
- Most people see adoptive parents as being most able to take care of children's needs across all dimensions.

Evaluation

This attitudinal survey importantly shows that negative normative beliefs about atypical parents persist, despite the existence of increasingly diverse family types.

The sample itself was relatively large and used a random sampling method, so it is representative. The use of a pilot study enhances the validity, as the researchers could ensure that the questions were measuring what they intended and could refine them if necessary. The survey would be repeatable and is therefore reliable.

The researchers suggested that future research in this area should also examine vignettes that incorporate more complex and intersectional demographic features, such as ethnicity and age.

CONTEMPORARY RESEARCH: CRIME: LITTERING, SOCIAL NORMS AND BROKEN WINDOWS THEORY

Study 1: Keizer, K., Lindenberg, S. and Steg, L. (2008). 'The Spreading of Disorder'. *Science*, 322(5908), pp. 1681–1685.

Study 2: Keuschnigg, M. and Wolbring, T. (2015). 'Disorder, social capital, and norm violation: Three field experiments on the broken windows thesis'. *Rationality and Society*, 27(1), pp. 96–126.

Study 1 overview

Broken windows theory (BWT) views crime as a rational activity, and suggests that people are more likely to break social norms if there is evidence that other people in the vicinity have also broken them.

Method and sample

Keizer et al. undertook a series of field experiments in the Netherlands to see whether people were more likely to litter if they believed that littering fit with the social norms of the area. The researchers tied flyers to cyclists' handlebars in two different areas where no bins were easily accessible, one alleyway covered in graffiti and one without graffiti. There were 77 research participants in each location.

Findings

They found that 69% of participants threw the flyer on the ground in the alleyway with graffiti but only 33% littered in the alleyway without graffiti.

Study 2 overview

Keuschnigg and Wolbring repeated Keizer et al.'s littering study in Germany, but added in an extra variable to test. They did not just want to know whether the local environment impacted littering, but also whether levels of social capital were correlated with littering.

Method and sample

They staged a field experiment in two different bicycle parking areas between university dormitories in areas of higher and lower social capital. To measure social capital, they distributed a survey to the student residents, which asked questions about whether the students felt connected to other people in the dormitory, how many friends they had in the dormitory, or if they cooked with others.

Again, they put flyers on the handlebars of students' bicycles in both areas between 4 and 5 a.m. To use the bicycles, the participants had to dispose of the flyers – either by littering or finding a bin.

The researchers conducted the experiment twice, once when they made no changes to the area (it was generally clean, without litter) and once, a week later, when they added litter (they emptied seven rubbish bags, 14 empty bottles and four cardboard boxes) to create 'disorder'. Weather conditions were the same on both occasions.

Findings

The table below summarises their findings (p. 106):

Social capital	Condition	Number of flyers prepared in the morning	Number of flyers still attached in the afternoon	Effective number of cases	Number of flyers properly disposed of	Number of flyers discarded to the ground
Low	Control	96	59	37	20 (54.05%)	17 (45.95%)
	Treatment	110	71	39	15 (38.46%)	24 (61.54%)
High	Control	81	43	38	29 (76.32%)	9 (23.68%)
	Treatment	99	49	50	26 (52.00%)	24 (48.00%)

Table 8.2: Number of cases across dormitories and experimental conditions. Source: Keuschnigg, M. and Wolbring, T. (2015). 'Disorder, social capital, and norm violation: Three field experiments on the broken windows thesis'.

More students dropped flyers on the floor in the area that was already highly littered ('treatment' in Table 8.2). The researchers also found that more students dropped flyers in areas where dormitories were deemed to be characterised by lower social capital. Therefore, levels of social capital and disorder independently impact student littering.

> **Question**
>
> What conclusions can you draw from this table?

Evaluation

Both studies provide empirical data to support the broken windows theory of crime. This is socially impactful, as police and law enforcement officials can act on this knowledge to reduce crime rates by tidying up neighbourhoods so as not to encourage law-breaking.

Field experiments tend to be more valid than laboratory experiments because they take place in natural settings. However, field experiments cannot control for all variables, so they are less reliable than other types of experiment. The method employed in these studies was only quasi-experimental, because it did not randomly assign participants to one group or another and the sample was not homogenised in terms of age and sex (variables shown to impact littering behaviour). The study also did not control for observer bias – the researchers were not blind to the studies' hypotheses, and their presence when observing may have impacted on whether or not littering took place.

Experiments as a social research method

1. Why are experiments the least used method in sociological research? What social factors might account for their decline in popularity?
2. Reflecting on some of the studies in this chapter, how successfully do you think ethical issues were mitigated?
3. Some sociologists integrate experimental questions into surveys. Is this really 'experimental' research and what insights does this approach to survey research afford?

Advantages of experimental research:

- Experimental research often provides researchers with a high level of control, particularly laboratory experiments.
- Laboratory experiments are said to have high levels of reliability, as they can be duplicated.
- Experimental research enables the testing of hypotheses and the establishment of correlations between variables.

Disadvantages of experimental research:

- It can be time-consuming.
- There are significant ethical issues – participants can be influenced and harmed.
- Laboratory experiments are artificial settings and therefore have low ecological validity, because they do not replicate the messiness and complexity of the social world.
- In field experiments, researchers cannot control for all variables, so these are less reliable than laboratory experiments.
- It is challenging to generalise from field experiments that manipulate behaviour in real-world settings.

CHAPTER 9: MIXED METHODS: SOCIETY AS A KALEIDOSCOPE

What are mixed methods?

As you have turned through the pages of this book, moving from chapter to chapter, you have also turned the social research kaleidoscope, each method revealing the different patterns, colours and contours of social life. The kaleidoscope analogy is helpful for social researchers, because when we turn the lens there are moments of clarity with distinct patterns, but also moments of blurriness, when the patterns are in movement and are less distinct. Moreover, it is not possible to reveal every permutation of the patterns all at once. This is also true of social life. The phenomena we study do not stand still for long – they are very complex. Partial insights are revealed but never the whole.

Sociology is, by its very nature, kaleidoscopic. Having different theories and methods means sociology is never stuck in one turn, revealing just one pattern, but instead exposes the social world in all its fabulous technicolour. Although this book has portrayed a neat and tidy overview of social research methods, focusing on one method and its insights in each chapter, it is important to remember two things:

- Sociology is open to a range of methods that, as Feyerabend noted, make it a good scientific discipline and give rich insights into different social problems. Social researchers refer to this as **methodological pluralism**.
- Some sociologists use multiple or 'mixed' methods to study a social problem or phenomenon. The use of mixed methods is the focus of this chapter.

> **Key concepts**
>
> **methodological pluralism**   the idea that using multiple methods gives multiple insights into social phenomena and allows a researcher to peel back layers, revealing more and more about the social world; this position is often held by realist sociologists

How do sociologists use mixed methods?

Sociologists who choose to use more than one method to understand a social problem may decide to draw those methods from the same methodological tradition – for example, two quantitative methods, such as surveys and official statistics, or two qualitative methods, such as ethnographic observations and interviews. Sometimes researchers might decide to cross methodological divisions and incorporate both quantitative and qualitative methods, such as combining surveys with interviews. In such cases, one of the approaches is usually more dominant than the other. For instance, social researchers conducting large surveys will often follow up with interview data; however, this is regarded as supplementary rather than of equal importance to the survey data.

Social researchers choose to use more than one method for three key reasons:

- They want to cross-check their data to enhance validity (triangulation).
- More than one method will help them answer their research question(s) more precisely.
- They take a realist approach to research and regard methodological pluralism as advantageous.

How rigorous are mixed methods?

Social researchers who advocate for the use of mixed methods argue that this approach is more rigorous than using only one method of inquiry. This is because mixed methods allow researchers to triangulate data. Triangulation is the process of cross-checking data to see whether the findings are valid. There are four main types of triangulation.

1. Data

This involves using a variety of data sources in a study, including time, space and people. For example, researchers might ensure that they gather data at different times of day, in different locations and with different groups of people. This is part of the sampling process.

2. Method

This involves using more than one method in a study. The methods may be drawn from the same or different research traditions (qualitative and quantitative).

3. Theory

This involves using more than one epistemological position in a study – for example, considering the research through positivist and realist or interpretivist and constructionist lenses (or combinations thereof). This type of triangulation is less common when an individual researcher undertakes a study by themselves, but is sometimes used in larger, collaborative studies, where several sociologists from different theoretical traditions come together to look at a social problem. Theoretical triangulation can be challenging, as different theoretical orientations may be considered incompatible. (See Thinking further below.)

4. Investigator

This involves using more than one researcher in the study to either gather data, interpret and analyse data, or both. The studies in this book that have used more than one interviewer or multiple researchers to analyse data and cross check interpretations are examples of triangulation by investigator.

Qualitative researchers using triangulation often argue that it enhances the study's credibility (the truthfulness criterion that replaces internal validity). Triangulation by investigator can also enhance a study's dependability (reliability), as other researchers act as auditors and check the research instruments – for example, the survey and/or methods of analysis – to ensure they are repeatable. Refer back to Chapter 1 to remind yourself of these alternative judgment criteria for qualitative research.

 Thinking further

Researchers' different theoretical orientations are sometimes described as paradigms (see Chapter 1). Typically, positivist research has been prioritised and afforded more credibility in the social sciences because its quantitative methods have been said to emulate the natural sciences, striving for objectivity and detachment. However, the emergence of qualitative methods underpinned by interpretivist paradigms, and the rise of social movements such as feminism, which have influenced realist and constructionist positions, has led to greater diversity in research.

Methodological pluralism has been pushed as an objective for sociology by some sociologists, as well as by governments and research councils that want students of sociology to be competent in a range of methodological techniques. However, fundamental theoretical tensions mean there are often clashes between social researchers concerning what constitutes 'good' and even 'proper' research. Some positivist researchers continue to argue that only 'scientific' inquiry that maintains objectivity and detachment constitutes good social research. Others argue that such 'scientific' approaches to research are incompatible with more critical and reflexive world views.

What sociologists are likely to use mixed methods?

In practice, many sociologists use multiple methods that might seem in contest with one another – at least theoretically. However, the main theoretical position advocating for mixed methods is the realist tradition, because realism aims to reconcile the differences between positivism and interpretivism – it sits somewhere in the middle. Realists agree with positivists that a social world exists independently of humans, and can therefore be studied. But realists also agree with interpretivists that we can only partially know this social world, through our own limited insights, so the aim of social research is to get as much knowledge and information about this external world as possible. Using lots of different research methods facilitates this.

The many lenses on housework

Throughout this book, you have read several research studies on housework, from classic research such as Oakley's interviews with women, to more contemporary studies that have examined the feminisation of 'zero waste' through document analyses. Go back through each chapter and create a mind map of the different methods used to examine housework. Then compare and contrast the different approaches and what each method reveals about this topic area. Consider the following questions:

- What insights were generated by the different studies into housework?
- Which do you think was the most effective study and why?
- Would any of the studies have benefited from using different methods?
- Do the findings from the different studies corroborate or challenge one another?

Researcher insights

A real value of mixed methods is that they enable a more careful and nuanced understanding. In a study designed to understand why people choose to pay for private health care, I (Sarah), with colleagues Mike Calnan and Jonathon Gabe (1993), used a survey and interviews. The survey asked questions about the experience of the National Health Service and people's levels of satisfaction with it. The data showed that people who had chosen to take out private health insurance were more dissatisfied with the NHS than those who did not have private health cover. However, the qualitative interviews revealed that this dissatisfaction did not mean that people wanted to opt out of the NHS and only have private health care. On the contrary, respondents expressed a strong commitment to state-funded health care. Without the interviews, this complexity would not have been discerned.

CLASSIC RESEARCH: WHY DO PEOPLE JOIN RELIGIOUS SECTS?

Barker, E. (1984). *The Making of a Moonie: Choice or Brainwashing?* Oxford: Blackwell.

Study overview

In the 1970s, there was much speculation and criticism of the Unification Church, which had been founded in Korea in 1954 and expanded into the USA under its leader, 'Moon'. The media claimed that this new religious movement was a bizarre sect and that its followers, known as 'Moonies', had been brainwashed. Barker, a British sociologist, wanted to know why people joined the sect and whether they had indeed been brainwashed.

Method and sample

To undertake this research, Barker employed multiple research methods, including interviews, observations and questionnaires. She spent nearly seven years on this research, and recounted how difficult it was to study the Moonies and to gain access. They were a closed group that had drawn a lot of criticism from the media, so they would be sceptical of researchers. Barker was not comfortable undertaking the study covertly – not least because she was concerned about the ethical issue of deception. She also felt that she would have aroused suspicion if she had tried to undertake her ethnographic work covertly, because she would have had to ask too many questions.

In the end, the Moonies themselves invited Barker to do the research because they wanted to set the record straight after all the criticism they had received. However, this posed a problem for Barker. While she was able to gain access to an otherwise closed group, she also wanted to ensure that there was no conflict of interest and that she could conduct her research objectively, without the bias of the group affecting the results.

Her three primary methods were interviews, participant observations and questionnaires.

To sample her interviewees, she secured a complete list of all the British Moonie members and then randomly selected 30 to interview. Each interview lasted 6–8 hours, during which she tried to find out about the members' backgrounds, why they joined the movement and their involvement in it.

Barker also secured access to the Moonie community to undertake participant observation over a six-year period. This involved living in different centres in England and in the USA, attending meetings for members and potential members.

In addition, she developed a 41-page questionnaire that she piloted with 20 Moonies and then distributed to around 500 other British and American members. This had a high response rate – only 11 people did not respond.

> **Questions**
>
> Why is covert participant observation regarded as unethical?
>
> How is Barker's perspective different to that of Humphreys, which you read about in Chapter 6?

> **Question**
>
> Barker uses various methods in her research. What are these methods and what are the advantages and disadvantages of using multiple methods to understand the Moonies?

Findings

Barker's research revealed that becoming a Moonie was a choice, not the product of brainwashing. Members had made the decision to join the Moonies themselves (they had agency), though some were more likely than others to join because of their own background and experiences.

Barker was quite open about the challenges of applying a scientific approach in the social sciences, but justifies it in the following way:

> For various reasons, some of which are fairly obvious, the subject matter of the social sciences is considerably less amenable to research than that of the natural sciences. But this does not mean that we cannot improve our knowledge of social patterns, trends and tendencies and gain a more reliable understanding of regularities between variables – of 'what goes with what'. By looking at groups as a whole we can begin to see patterns of relationships, and it becomes easier to detect which occurrences are incidental, even if sensational, and which are 'normal'. (p. 26)

Evaluation

Barker's research was clearly sociologically interesting, as it gave insight into an otherwise secretive part of society. It was also important because Barker dispelled several enduring and discriminatory stereotypes about the Moonies. (You can draw comparisons here to Humphreys' ethnographic research.)

It would be challenging to repeat the study, so it scores low on reliability. In contrast, it is highly valid, as Barker's methodological pluralism enabled her to gain a comprehensive understanding of the Moonies. She gathered quantitative data, as well as speaking to and observing the behaviour of group members in their natural settings.

Ethically, there are some issues with researching a group that is typically closed off from wider society and which therefore may contain some sensitive information. However, Barker chose to be overt in her observational research role, and therefore respondents could give their informed consent to participate. Fully determining whether consent was given willingly or unwillingly, however, is challenging because the Moonies had asked Barker to undertake the research, so some participants may have felt compelled to talk to her as the group leaders had authorised it. Barker felt that this outsider status was beneficial on the whole, though, as her experience revealed that many insiders opened up more to her because of her detached observer role and she developed an empathetic understanding – *Verstehen* – of those she studied.

> **Questions**
>
> Barker claims that social scientific research should 'improve our knowledge of social patterns, trends, and tendencies'.
>
> What theoretical perspective do you think she comes from?
>
> What research method would you usually expect to see from someone with this theoretical lens?
>
> What contradictions can you identify in her use of different research methods?

CONTEMPORARY RESEARCH: FAMILY: THE FOURTH SHIFT

Venn, S., Arber, S., Meadows, R. and Hislop, J. (2008). 'The fourth shift: exploring the gendered nature of sleep disruption among couples with children'. *The British Journal of Sociology*, 59(1), pp. 79–97.

Study overview

This study was designed to examine the extent to which having children disrupts sleep and whether this disruption varies by gender. The researchers observed that studying sleep is incredibly difficult – not only does it happen in the privacy of people's homes, but it is a largely unconscious activity, which makes it more difficult for people to talk about. The researchers therefore argued that to study sleep they needed mixed – and innovative – methods.

Method and sample

The researchers used interviews with couples, audio sleep diaries, and individual interviews to follow up on the audio sleep diaries. They used a convenience and snowball sample, and attempted to recruit different socio-economic groups, although all respondents were in heterosexual couples with children. The respondents were between 20 and 59 years old and were at different stages in their lives with children. Eighteen interviews were conducted with couples whose children lived at home and seven with couples whose children had moved away but returned home for extended visits.

Topics included in the couples' interview guide were the quality, quantity and nature of sleep, sleep expectations versus sleep reality, external influences on sleep, including children, other caring responsibilities, pets, etc. The researchers also asked about physiological factors, including needing to go to the toilet, snoring and pain.

The researchers used a second method of one week's worth of daily audio sleep diaries (a documentary method). Respondents were given a hand-held recording device and instructions to talk about their sleep within 20 minutes of waking. They were given loose guidelines of the kind of things they might talk about, but were not required to complete structured sleep logs. In this sense, the data captured was mainly qualitative, although at the end of each log participants were asked to rate their night's sleep on a scale of 1–5, with 1 being 'very poor sleep' and 5 being 'very good sleep'. This enabled a comparison across nights of the week.

Finally, the researchers conducted individual in-depth interviews which followed up on the sleep diary content. The data from all three methods was then thematically analysed to formulate theories.

> **Question**
>
> What do you think might be some advantages and disadvantages of using audio sleep diaries as a method to capture data on the domestic division of sleep?

> **Questions**
>
> Why do you think the researchers chose to interview the participants a) in their couples and b) individually?
>
> What are the advantages and disadvantages of both approaches to interviewing?

Findings

Below are some excerpts from the sleep diaries.

> 40F(AD) [Youngest son] woke up at five past four and it took ten minutes to get him to settle and I went back to bed at four-fifteen. [Oldest son] woke up at twenty past five and came in bed with us and, umm, and woke up at about five to seven, but we actually got up at twenty-five past seven, so we dozed for a little while, umm, talked to [oldest son] a little bit. (Two sons, 5 and 2½ years)
>
> 40M(AD) But umm, I don't recall [youngest son] or [oldest son] waking me up during the night. Oh yes, actually [oldest son] did come in didn't he, telling us about he had a bad dream or something, I can't remember what time that was either. (p. 94)

Significantly, the researchers found evidence of a 'fourth shift' for women. Building on the second shift (childcare) and **third shift** (emotional work), the fourth shift is a 'night shift' that is predominantly undertaken by women, regardless of whether they are in employment or not. The figure from the study illustrates this concept.

Men did undertake some night-time responsibilities for children, but this tended to be when the children were older and were staying out late. In these cases, men assumed a role as the family 'protector'.

First shift
Day time work
For example, paid employment or childcare and domestic chores

Third shift
Emotional activity: Throughout first and second shifts, thinking about needs of children, partner, employment

Second shift
Undertaken in the evening
For example, domestic chores, putting children to bed

Fourth shift – night time
- Emotional activity at night related to childen, work, partners, and other family members
- Getting up in the night to deal with children's needs or domestic chores

Figure 9.1: The fourth shift. Source: Venn, S., Arber, S., Meadows, R. and Hislop, J. (2008). 'The fourth shift: exploring the gendered nature of sleep disruption among couples with children' (p. 81).

Evaluation

The researchers importantly established a new theory that built on previous theorisations of the domestic division of labour, and introduced the concept of the 'fourth shift'.

While the researchers tried to include a diverse sample across social classes, the study was not overt about the ethnicity of the participants and was also homogenised to heterosexual couples. In addition, the study had a relatively small sample size, which limits its representativeness. The study's validity is enhanced by methodological triangulation.

Researchers carried out initial interviews with couples and then completed follow-up interviews with individuals. However, the validity of the data from the first interview could be questioned, as it is possible that the husbands' presence at the interview influenced the women's responses, and that the women may not have spoken as freely as they would if they were interviewed individually.

CONTEMPORARY RESEARCH: CRIME: MORAL PANICS OVER DRINK 'SPIKING'

Burgess, A., Donovan, P. and Moore, S. (2009). Embodying uncertainty?: Understanding heightened risk perception of drink "spiking"'. *British Journal of Criminology*, 49(6), pp. 848–862.

Study overview

Research by Burgess et al. used interview and survey methods to understand women's perceptions of the risk of drug-facilitated sexual assault (DFSA) through what is commonly known as 'drink spiking.' 'Date rape' has become something of a moral panic spread through the media to refer to this type of DFSA crime. The researchers were interested in finding out about the nature, extent and basis of students' fears of sexual assault from drink spiking. They argued that a 'sociological imagination' was needed to understand the relationship between perception and social reality.

Method and sample

Burgess et al. administered 200 questionnaires at three UK universities (Kent, Sussex and London) during 2006–07. Through the survey, they aimed to gather data on students' knowledge of date-rape drugs, their association with victims of these drugs, and whether students' behaviour had changed in relation to the perceived threat. The survey replicated a previous study undertaken in the USA.

The survey data was supplemented by qualitative data from 20 in-depth interviews and four focus groups. Most interviewees were female (85.3%), second-year undergraduates, with a mean age of 21.

The researchers used a convenience sample, in which surveys were distributed after lectures and at campus cafés across the three participating universities. The interview sample was also opportunistic: an advert was placed on university web pages to recruit participants.

Findings

The quantitative data revealed that few people had actually experienced drink spiking, despite high levels of awareness of it and a high sense of perceived threat.

Ten respondents had personally experienced drink-spiking, and none had been subject to sexual assault as a result. Most respondents, however, were aware of the threat of date rape drugs (95% UK respondents) and 55% of respondents claimed to know someone who had had their drink spiked.

> **Question**
>
> Look at the data in Table 9.1 on respondents' perceptions of risk from sexual assault. What does this quantitative data reveal?

	Number and percentage that identified the given risk factors
When drunk	144 (72%)
When walking alone at night	140 (70%)
After having taken drugs	44 (22%)
After having your drink spiked with drugs	150 (75%)
In your home	10 (5%)
When walking in an area where crime is known to be high	58 (29%)

Table 9.1: Responses to the question: 'Under what circumstances do you consider yourself to be most at risk from sexual assault (please tick no more than three boxes)'.

The results show that 75% of respondents said that having a drink spiked was the biggest risk factor for sexual assault – more significant than drinking alcohol or taking drugs.

Interviews with respondents elicited further details on perceived risks. Below is a quotation from an interview with 'Freshers' (first-year university students) about their perceived risks during Freshers Week.

Interviewer: How was Fresher's Week?

Interviewee: I tried to stay in during Fresher's Week, 'cos there's only a few people I know properly here, and I really wanted to get to know people before we went out … I've been out with my house mates and I've only been off campus once … I like not so big groups, because you know where everyone is then.

Interviewer: So going out is risky?

Interviewee: Yeah ... like, I don't even like going to the toilet on my own. I went out last night [to a local nightclub], and two guys came up to me, talking as if they knew me.

Interviewee: I'm always quite precautionary, 'cos when you're drunk you've got to be, you can't ... you've always got to be aware of things. But I think there is always [the threat of drink-spiking] in the back of your mind now. And I don't necessarily think that's a bad thing, it just means you're more cautious ... it's a slight paranoia. I mean, I wouldn't get paranoid to the extent that it ruins a night, but it's a concern. Especially because I don't really drink bottled drinks ... I mean, you don't always want to carry [your drink] around everywhere but, really, that's inevitable ... Like, before, in the Venue you could leave your drinks and go for a dance. But not so much now. I only leave [my drink] with friends, but even then ... [hesitates]. (p. 854)

The interview data provides a more rounded picture of perceptions of risk. It does not just reveal that students regard drink spiking as a significant risk; it also shows the kinds of mitigating behaviours that young women engage in to try to ensure it does not happen to them.

Evaluation

The study focuses on a socially important topic – that of drink spiking – and reveals that although perception of the risk of sexual assault from drink spiking was high among British university students, the likelihood of having one's drink spiked was very low. Thus, the authors suggest there is a moral panic around drink spiking, and as a result young women engage in lots of mitigating behaviours to avoid it.

The survey instrument was arguably reliable, as it had been tested in a previous survey in the USA. Because there was a large dataset from that earlier study, the UK data could be compared to it, increasing the study's reliability.

The sample is quite small, consisting of only 200 questionnaires, and with students from only three universities. The use of convenience sampling also means that the results are not fully representative of the wider student population. The interview sample is self-selecting, and therefore the representativeness can be questioned. However, the mixed methods design arguably increases the study's validity (methodological triangulation), as it provides a more comprehensive account of young women's perceptions of date rape and drink spiking.

Questions

Why do you think the researchers used quantitative and qualitative methods in this study? What are the advantages of using more than one method to understand a topic like drink spiking?

CONTEMPORARY RESEARCH: EDUCATION: SCHOOL 'BOFFINS'

Francis, B. (2009). 'The Role of the Boffin as Abject other in Gendered Performances of School Achievement'. *The Sociological Review*, 57(4), pp. 654–669.

Study overview

Francis's study looks at the stereotyping of high-achieving students as 'Boffins' and how this label is given to, and performed by, school children. She was particularly interested in high-achieving students because the discourses that circulate around 'underachievers' also tend to be related to how popularity (in relation to peers) might impact educational achievement. Is it possible to be popular *and* high-achieving? If high achievers are labelled negatively, does this hold them back for fear of stigmatisation?

Method and sample

Francis studied high-achieving Year 8 students (12–13-year-olds) through two main methods: ethnographic observations and 71 interviews (36 girls and 35 boys). Both methods were conducted in nine different state secondary schools across different regions in England. Ten students were from minority ethnic backgrounds. Students were also classified according to their social class, which was indicated by their parents' occupations and their residence (postcode). In total, 39 students were classified as middle class, 23 working class, and nine remained uncategorised.

'High achievement' had to be operationalised: teachers identified high-achieving students and then Francis measured this achievement by considering the students' educational credentials, including whether students were in top sets and streams across a wide range of subjects. Evidence to support this came from their Key Stage 2 Standard Assessment Test (SAT) results and recent subject grades. The notion of high achievement, however, differed from school to school.

Another concept that Francis had to operationalise was 'popularity', but she also recognised this would be tricky to measure:

> 'Popularity' is clearly a complex and slippery concept, both in its actual meaning (those most popular are not necessarily those most liked as the concept incorporates aspects such as influence and admiration) and in perspective (those most popular with some groups may not be popular with others). We were interested in pupil popularity among peers, rather than with teachers, as it is gendered peer-group power and status relations which are argued in the literature cited above to have an impact on gender and achievement. Hence all pupils in identified top stream classes were asked to complete a short survey on popularity in their class. (p. 648)

Questions

What challenges did Francis find with regards to operationalising the concepts of 'high achievement' and 'popularity'?

Do you think she did this well? What issues did she face? How else could you operationalise these concepts?

The notion of 'high achievement' differed from school to school. Does this raise any issues with regards to the study's reliability?

Francis conducted ethnographic observations in each school in order to document classroom interactions. She focused on the high-achieving students' behaviour. She tracked classes throughout different lessons for a full school day and took additional notes during breaks and lunchtimes. Her interviews were also conducted in schools.

Findings

Francis conducted a qualitative thematic analysis through a social constructionist lens. She found that the label 'Boffin' was a stigmatised term used in a derogatory way. It was constructed in relation to various themes:

- **The Boffin as a 'social pariah':** The label signifies 'isolation and social rejection' – students did not want to be seen as a Boffin because they regarded it as an 'oppressive' identity.
- **A lack of 'balance':** The Boffin label was applied to those who were seen as 'unbalanced' and who focused on educational success at the expense of more rounded educational and social experiences.
- **Asexuality or homosexuality:** The Boffin was constructed by students in association with sexuality, where high-achieving boys ('Swots') were often labelled as 'effeminate' and not expressing normalised 'laddish' hegemonic masculinity.

Some students were able to manage the Boffin label more readily than others, but this depended on the other aspects of their identities that affected popularity. This management typically happened in two ways.

- **'Jealousy':** Some high-achieving students would manage the Boffin label by challenging those who labelled them Boffins and bullied them or other students for being high achievers. These students saw bullies as 'jealous' of their educational success. High-achieving students would claim 'superiority' to justify this. Sometimes, high-achieving students would draw on wider discourses of social class and would also apply derogatory labels to students they regarded as educationally inferior to them as 'chavs'.

> **Questions**
>
> What type of observations did Francis use (for example, complete observer, participant observer) and why?
>
> Why do you think Francis chose to conduct interviews as well as observations? What advantages are there of using the two methods combined, rather than using one or the other?

- **'Girling':** Francis notes that high-achieving girls would often manage the Boffin label by drawing on practices of girling (emphasising their femininity and attractiveness) to resist the label and maintain popularity.

Boys were less able than girls to manage the 'Boffin' label. The discourse of hegemonic masculinity constrained boys, as it also intersected with sexuality to label boys as 'gay' for being high achievers.

Below are some excerpts from Francis's interview data.

> Helen: if you're a girl then you can get away with trying hard and doing well in your subjects, but if you're a boy then you can't get away with –. You can get away with being smart but you can't get away with, really trying hard. [...] People'd just think you was a bit of a, gay, if you did, if you really tried.

> Stella (white, middle-class, Ironoaks): [...] there's a boy in the year above us who is like really camp and his name's Josh [...] and he, he got really badly bullied last year, people used to like beat him up in the changing rooms and throw away his clothes and stuff [...] But um yeh he used to like get bullied really bad, because he wasn't, because he was clever and he didn't, you know, sort of act all macho. (p. 662)

Francis's study shows that boys are often labelled as 'underachievers' in educational settings. However, these labels are socially produced by wider discourses of hegemonic masculinity, sexuality and social class, which shape boys' behaviour in the classroom and constrain their educational successes.

Questions

What are the different insights gleaned from the interviews and the observations in this study?

Do you think that Francis needed to use both methods? Why, or why not?

Evaluation

Francis's study is clearly sociologically important. It reveals why some students might intentionally underperform in order to fit in with wider social norms concerning gender and sexuality in educational settings, to avoid being labelled as a 'Boffin'. Students who want to be popular at school might therefore choose to hide their educational successes.

The sample was relatively large and diverse according to gender, social class and ethnicity, and was conducted across different regions in the UK. This makes it representative, although, as the sampling method was not random, it was not generalisable to the wider UK population.

Because the researcher used observations and interviews, the study has validity – observed behaviour could be cross-checked through interview data. More importantly, the combination of methods gave a more nuanced understanding of the way that 'Boffinhood' is managed.

It would be difficult to repeat this study and so the results are less reliable.

Case study: Marxism and the impact of social class on identity in schools

Francis's study showed the impact of identity on educational achievement. In a classic study, *Learning to Labour*, Willis (1978/2016) took a Marxist perspective and used mixed methods (case studies, participant observations, interviews and focus groups) to unpack the way that schools socialised working-class boys, or 'lads', into working-class jobs. Willis revealed the hidden structure of capitalism that limited working-class social aspirations and, in turn, social mobility.

Willis identified two groups in the school – the 'lads' and the 'ear-oles' (similar to Francis's Boffins). The working-class 'lads' expressed negativity towards school and learning, they believed there to be few opportunities in their future, and recognised that middle-class children had better chances. In this way, their fatalistic approach accepted that meritocracy was a myth, manifesting itself in not taking school seriously; instead they focused their energy on larking around. In contrast, the 'ear-oles' embraced the ethos of learning, worked hard and respected their teachers. Willis showed how these school subcultures effectively reproduced class differences.

This was an extremely influential piece of work. Willis conducted 18 months of observations in the school setting, and followed the boys into work for a further six months. However, the data was only drawn from the experiences of a sample of 12 White students from one school in Birmingham.

Mixed methods in social research

1. Why do you think not all researchers choose to use mixed methods?
2. Review the studies in this chapter. Which were most successful in triangulating the research and why?
3. What additional insights did mixed methods afford in the studies described in this chapter?

Advantages of mixed methods:

- Using mixed methods gives a deeper and more comprehensive account of a social phenomenon.
- Researchers can be more confident in the findings when methods are triangulated, so mixed-method studies have higher validity than those using only one method.
- Mixed methods can balance out the limitations of the different methods and epistemological positions underpinning them.
- Researchers can triangulate by investigator and increase the validity and reliability of the study.

Disadvantages of mixed methods:

- Paradigm clashes/theoretical contradictions are hard to reconcile when using different methods that seem at odds with one another.
- Mixed-method research is arguably less reliable if conducted by one person (when investigator triangulation does not take place) because it is more challenging to repeat in exactly the same way, particularly if it comprises ethnographic fieldwork or interviews as one of the methods.
- The approach is more time-consuming and complex to carry out, and can be more costly.
- It requires expertise in different methods.

CHAPTER 10: SHAKING UP METHODS

In this chapter, we move from 'mixing' methods to 'shaking up' methods. In recent years there has been a more critical and reflexive turn within sociology. This has tended to emanate from constructionist, postmodernist and postcolonial (and decolonial) theoretical positions, and is sometimes associated with a new set of methods. Sociologists advocating that methods should be shaken up argue that there are other epistemologies that shape *what* we can know about the social world and *how* we can know about it through research methods.

As critical and reflexive researchers, sociologists who want to shake things up are mindful of the historical and geographical located-ness of research methods, and how research methods are both products and conduits of unequal power relations. As such, they recognise the importance of devising new and innovative methods that are timely and ethical. For example, sociologists aim to 'decolonise' and 'queer' methods, reflecting on the Eurocentrism, heteronormativity, cisnormativity and **anthropocentrism** of many traditional methods. As a response, they include more embodied and reflexive ways of knowing.

This chapter explores five examples of new methods used to shake up sociological approaches, and their underpinning epistemologies.

Autoethnographic and ethnobiographic methods

Traditionally, sociologists have cautioned against drawing upon personal and biographical experience as data. Indeed, Mills (1959/2000) and Bauman (1990) both argue that a key methodological rule for sociologists is to move beyond individual experience, ensuring that sample sizes are large enough to eradicate personal bias and skewed findings. However, more recently, sociologists have recognised the insights that autoethnographic and ethnobiographic research can afford. These methods do not abandon the sociological project to connect personal experience to public social structures; indeed, they set personal experiences alongside or in the context of wider cultural reflections.

Autoethnographies tend to reflexively document the experiences of the researcher themselves (autobiography) and set these in a wider social context (ethnography). Ethnobiographic methods draw on experiences of another person (biography) and subject this to sociological analysis through ethnographic reflection. Both approaches use inductive methods to generate interpretive accounts of social life. Sometimes, researchers in these traditions are drawn to social constructionist and postmodern perspectives, and aim to reflexively consider the power relations that exist between themselves as researchers, their research subjects and their own life experiences that shape the study. (See the Contemporary study 'Becoming a skinhead' below.)

Embodied methods

Some sociologists have argued that research methods require ways of capturing more embodied forms of data – that is, knowledge about (and experienced through) the body. 'Embodiment' describes how people experience their bodies as subjects rather than as objects. Historically, there has existed a perceived **dualism** between mind and body – that is, we use our minds to 'know' about our bodies. Embodiment draws from a **phenomenological** position known as **monism** to argue that the mind and body are not two separate entities, but that we are instead one entity – mind and body together, embodied. This approach argues that knowledge about the social world is shaped by and through physical bodies, and that physical bodies are shaped by society. Sociologists drawing on the work of Bourdieu refer to 'bodily habitus' – ways of acting and being that are contained within our bodily form and movement, and are shaped by access to resources. (See the Contemporary study on pregnancy and 'leaky bodies' below.)

Post-qualitative and post-human methods

Sociologists have also called for methods that consider the changing nature of the human body itself. Post-qualitative methods draw from theories of postmodernism and **post-humanism**. At its core these focus on understanding relations between humans and non-humans. For example, Haraway (2006) introduced the important concept of 'cyborg' to capture the ways that technologies and physical bodies have become intimately entwined – consider pacemakers, cochlear implants, artificial limbs and contact lenses, for example. Lupton (2019) has also researched the new relationships between what she calls humans and 'data selves', outlined in her study in this chapter. More than this, sociologists like Smart (2011) challenge the anthropocentrism of sociology and demand that sociologists incorporate research that examines relationships between humans and animals.

Thinking further

Shilling (2012) argued that sociologists must incorporate the 'lived body' into their research. However, this raises several challenging questions for social researchers. In particular:

- how can sociologists research the physical 'body' and people's bodily experiences?
- How can sociologists 'know' about people's lived experiences if the only way of explaining it is through language?

When sociologists turn fleshy, lived experiences into words, there are inevitably aspects of that experience that cannot be fully grasped. For example, try explaining to someone what it is like to go out for a run and be out of breath, or exactly what it is like to eat your breakfast. Imagine what it must be like to experience childbirth (as the women in Lupton's study do). Sociologists writing about the body, therefore, may have problems fully accessing and representing other people's experiences (consider this in relation to the 'crisis of representation' in Chapter 1).

Case study: Canny knowledge and crossing species boundaries

In her study 'Ways of Knowing: crossing species boundaries', Smart (2011) uses observations and unstructured interviews with horse racers to try to understand their relationships with their horses. She acknowledges the difficulties of using words through interviews to describe experiences like human–horse interactions, which are, to some extent, 'beyond words'. She uses the concept of canny knowledge (knowledge that is acquired experientially) versus uncanny knowledge (evidence-based knowing) to explain people's relationship with their horses. Smart recalls in her interviews, for instance, how participants used bodily gestures to explain elements of horse riding, with one participant taking her to the stables to demonstrate through movement that which she could not describe in words.

Case study: 'Post-qualitative' research: Cyborgs and data selves

Lupton's book *Data Selves: More-than-Human Perspectives* (2019) revealed the new ways that humans interact with technology, showing that our bodies and our selves are intertwined with data. Lupton and Watson (2021) used interactive workshops to creatively research how people interact with personal data held through technology (for example, phones and the cloud). In one workshop activity, called Data Letters, participants were invited to write a love letter or breakup letter to their personal data. In another activity, Data Kondo, they used TV presenter Marie Kondo's principle of 'sparking joy' to ask participants to focus on the research questions 'How do we feel about our personal data and how do we curate our personal data, deciding what to keep and what to discard?' Participants were asked to respond to the following prompt:

> *Imagine that you are clearing out and de-cluttering your digital device, as you would your house or apartment, based on Marie Kondo's principles. You are getting rid of data clutter (your images, videos, music, documents, steps taken and any other information about you). What personal data would 'spark joy' for you? What data wouldn't, so that you are happy to get rid of them? Who would you give your discarded data to – and why? Or would you just delete your data permanently? (p. 472).*

The idea of these workshops was to use innovative and arts-based methods to gather information on how data selves and identities are constructed in new ways. Lupton described this as an example of what other social researchers have called 'post-qualitative inquiry'. Key to this is the centring of the relationship between humans and non-human **assemblages** – that is:

> *[…] human researchers and participants or co-researchers with things such as words, recording devices, pen and paper. (p. 473)*

Decolonial methods

Increasingly, sociologists recognise that social research methodology is largely Eurocentric. This means that it ignores and undermines Global Southern knowledge and especially that of indigenous and marginalised social groups, privileging White westernised ways of knowing. To decolonise research methodologies, researchers must ask how power relations affect different stages in the research process. In particular, it is important to assess:

- the types of methods that have traditionally been used
- the types of questions that have traditionally been asked
- the types of literature that are traditionally drawn upon
- who traditionally undertakes the research
- who funds the research
- who is the subject (or object) of the research.

An example of a decolonised method is given in the research study below, 'The same as everyone else', which explores the impact of the hidden curriculum on the Māori indigenous population of New Zealand.

Queering methods

Social researchers have also stressed the need to 'queer' methods (Compton et al., 2018; Ghaziani and Brim, 2019). 'Queer' is used here to mean destabilising classificatory systems and disrupting assumptions, particularly concerning sexualities and genders. A queer perspective emphasises the fluidity and multiplicity of identities in a society where dominant normative identity categories prevail. Queer studies therefore incorporate **transgender** studies, **fat studies**, **critical disability studies** and critical heterosexualities, as well as **quare studies**, which encapsulate racialised sexual knowledge. Queer theory regards the social world as 'vague, diffuse, or unspecific, slippery, emotional, ephemeral, illusive, or indistinct, changes like a kaleidoscope, or doesn't really have much of a pattern at all' (Law, 2004, cited in Ghaziani and Brim, 2019, p. 13).

Queer sociologists argue that sociological knowledge has been blinkered to, and has subjugated, diverse lived experiences, and in doing so has reinforced rather than challenged normative assumptions about genders and sexualities. Considering that traditional social research methods aim to impose classifications in order to collect and analyse data, its practice is challenging for queer theory. Because queer theory rejects the notion of unchanging, impermeable categories and dualisms, existing research methods cannot capture the complexity of the social world as they describe it. For example, survey questions that gather demographic data will ask respondents to choose between a number of discrete choices (for example, male, female, other). However, such categorisation is at odds with the agenda of queer theory, which is critical of such categories in the first place and argues that gender identity is fluid rather than fixed.

Positivist social research methods are therefore often not regarded as fit for purpose for understanding **LGBTQIA+** communities. Queer social researchers are influenced by constructionist and postmodernist epistemologies, always aiming to destabilise categories, and seeing identities as multiple and intersecting.

Methods are queered when researchers use queer theory to reimagine or disrupt what is possible with existing methodological techniques. Queer research methods embrace the complexity of the world, so queer researchers either have to adapt existing methods or develop new methods of inquiry. Plummer (2005) argues that researchers can use subversive ethnographies to challenge heteronormative assumptions, or they could engage in 'scavenger methods', which mix methods that are often cast as being at odds with one another.

Case study: Queer scavenger methods

Halberstam's 1998 study on female masculinity drew on a range of methods, including ethnographic film research, archival documentary analysis and historical surveys, as well as literary textual analysis. It explored myriad masculinities of women ranging from, but not exclusively encompassing, aristocratic European cross-dressing in the 1920s, tomboys, butch lesbians and dykes, drag kings, and trans-butches and transsexuals (female to male). Halberstam's scavenger method aimed to draw attention to performances of female masculinity that had traditionally been omitted from sociological enquiry. Her study decoupled gender from sex, and masculinity from men. Instead of focusing on the dominant White male masculine narrative, which she regarded as uninteresting, she sidestepped the stigma of masculine women and made them the centre of her analysis.

CONTEMPORARY RESEARCH: CRIME: ETHNOBIOGRAPHY – BECOMING A SKINHEAD

Beauchez, J. (2021). 'Becoming a skinhead: An ethnobiography of brutalized life and reflective violence'. *The Sociological Review*, 69(6), pp. 1179–1194.

Study overview

Beauchez wanted to study deviance and crime in the skinhead subculture. To do this, he focused on the life of one skinhead, Yvan, who Beauchez knew from his childhood.

The skinhead subculture emerged in England in the 1960s but spread to other parts of the globe, including France, where Yvan and Beauchez resided. It was a predominantly White working-class subculture associated with particular clothing, including Doc Marten and steel-capped boots, and the 'skin' or shaved head. While the movement was often associated with antiracism, because of its links with reggae and other Black Caribbean influences, it was also embedded with **xenophobia** and closure towards 'others'.

For example, although there were different strands of the skinhead subculture, one of these strands in the 1990s was particularly associated with neo-Nazism and racism. Yvan himself was taken in by this nationalist skinhead strand at the age of 12, and Beauchez wanted to explore his socialisation into this group in relation to his family background, social class and wider subcultural group affiliations.

Method and sample

Beauchez focused on the life of one person – Yvan – who died aged 28. It is an interesting sociological decision to focus on only one respondent. Beauchez grew up with Yvan and claimed to have a unique 'insider perspective'. He therefore reflected on his own researcher socialisation alongside documenting his respondent's life. He wanted to know what it meant to live as an 'outsider' – as someone associated with the skinhead subculture, which was despised by many members of society.

Biographical approaches have been used in criminological and sociological studies to understand experiences of deviancy and to document the stories of people who live 'on the margins'. Sociologists using this method argue that such in-depth accounts of one individual enable the researcher to understand the 'criminal self'. However, as with all good sociological methods, the focus is not simply on understanding the self in isolation, but also on how it is structured and discriminated against by wider social forces.

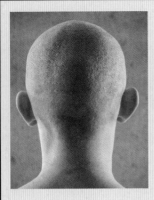

Ethnobiography combines this storytelling with experience of the field as an insider. Beauchez interviewed Yvan for 15 hours, spread over four separate occasions, about his life as a skinhead. The interviews focused on 'reliving our shared and separate experiences of the social world and its margins' (p. 1183). As Beauchez wrote in defence of his methodological choice: 'It is the depth of this shared experience that lends ethnobiographical value to these interviews insofar as the events recounted were first experienced, and often shared.' (p. 1183)

Findings

Beauchez's interview and ethnographic data were written up in a story-like way, documenting Yvan's pathway in and out of different subcultural groups. Beauchez details through thick description how Yvan was adopted as a child into a working-class French family that moved from the council estates into the suburbs, shaped by middle-class ambitions and an othering of 'immigrants' who moved into the council houses they had once occupied and 'escaped' from.

Beauchez's analysis is informed by Bourdieu's work on distinction – Yvan's family, like many others, aspired to be like the higher classes, in polite society, and wanted to distinguish themselves from working-class culture. Yvan did not meet his father's standards. He suffered violence at his father's hands, and as a consequence they had a mutual contempt for one another:

> I was twelve when I got tattooed and thirteen when I went smooth [shaved my head], to stand out, you know. [...] And it was a good way of pissing off the old man as well: looking like everything he hated the most. [...] And actually, as soon as he saw me with my shaved head, he said I was a poof, a sissy, an Auschwitz survivor, and all the rest. [...] You know, the usual: for him, I was just a piece of shit. (p. 1185)

Becoming an outsider, Yvan started hanging out at a local Gypsy-Traveller camp, where he acquired a substitute mother who was later brutally murdered. There, he engaged in consensual tattooing – historical markers of criminality – including a 'white lightning' tattoo (symbolising the military branch of the Nazi regime). With his shaved head and tattooed body, Yvan was soon recruited to a local skinhead group.

Beauchez's data reveals how Yvan did not necessarily like or agree with the racism and violent nationalism of the group, but he was comforted by the solidarity (swear words are redacted):

> No one could stand the skins. Which was good: '[...] the world!' [in English] That was one of the first tattoos I ever got. But you know, those guys are some of the only people I've ever been able to count on. It's with them that I felt the most solidarity. (p. 1188)

Yvan socialised with children from poor backgrounds who had dropped out of or had been expelled from school. Petty crime and neutralised aggression (crimes towards people they saw as 'fair game', such as sexually predatory men) relieved boredom and gave meaning to Yvan's life.

Beauchez coined the term 'reflective violence' to describe the violence Yvan and other skinheads both experience and commit as a reflection of the wider social contempt that they receive. They are labelled as troublemakers and they reflect this violence back to society, marginalising themselves even further.

Yvan's friend Vince, who joined him for these criminal escapades, was imprisoned. On his release he re-joined and headed up the skinhead group, which became even more nationalistic and racist. Vince acquired new tattoos that displayed his racism and Nazi affiliation more prominently, and Yvan was ousted from the group for his 'mixed' identity, which affiliated with Gypsy-Travellers, and because he engaged in drug-taking (something seen as not fitting with the Nazi strand of 'skins'). As a result, Yvan began connecting with another division in the subculture – one with a more mixed marginality. He also decorated his body with further displays rejecting Nazism, including a tattoo that read: 'I'm not racist, I hate everyone.'

Through the narrative of Yvan's life, Beauchez depicted him as a complex character, and showed the equally complex nature of the different strands of the skinhead subculture.

> **Questions**
>
> How representative is Beauchez's research into deviance and subcultures?
>
> What are the advantages and disadvantages of the ethnobiographical method used here?
>
> How valid and reliable do you think this data is?

Evaluation

Because Beauchez interviewed one respondent (Yvan), and supplemented this with observations in his natural setting, the data might be said to be valid – that is, it is likely to be accurate and truthful. However, it is not necessarily reliable because it would be impossible to repeat this study with either the same respondent (Yvan passed away) or with another person experiencing life in the skinhead subculture (as their own life course would be very different).

The method is also potentially problematic from a deeper sociological perspective. Mills (1959/2000) reminds us that personal troubles (or truths or triumphs) are socially structured and can only be regarded as social problems if they are *shared* by a wider social collective:

> *When, in a city of 100,000, only one man is unemployed, that is his personal trouble, and for its relief we properly look to the character of the man, his skills, and his immediate opportunities. But, when in a nation of 50 million employees, 15 million are unemployed, that is an issue, and we may not hope to find its solution within a range of opportunities open to any one individual. (p. 9)*

Is Beuchez's ethnobiographical method properly sociological and ethical? Despite only focusing on one person's life, the data gives an emic view – an insider perspective that is rich and detailed. In this way it can be regarded as sociological. Researchers using ethnobiographical methods might find it hard to protect the identity of the respondent and maintain confidentiality and anonymity (as it could be easy to guess who the study is referring to since it documents their life). This raises questions of how ethical the study might be.

> **Question**
>
> What other ethical issues might be raised by this ethnobiographical method?

CONTEMPORARY RESEARCH: FAMILY: EMBODIMENT, 'LEAKY BODIES' AND CHILDBIRTH

Lupton, D. and Schmeid, V. (2013). 'Splitting bodies/selves': Women's concepts of embodiment at the moment of birth. *Sociology of Health & Illness*, 35(6), pp. 828–841.

Study overview

Lupton and Schmeid wanted to understand women's experiences of childbirth from an embodied rather than a detached and objective perspective.

The concept of embodiment challenges the traditional dualism that privileges the mind over the body. This dualism has historically been used to create further dualisms between men and women. Men (particularly White men) are constructed as rational and in alignment with the mind. Women are often framed as emotional – driven by their bodies. Feminist researcher Shildrick (1997) coined the term 'leaky body' to describe how women's bodies are often seen as unregulated, uncontained and excessive – in effect, 'leaky' – in patriarchal society.

Lupton and Schmeid argued that dominant discourses uncritically construct the pregnant and labouring body as leaky – ruled by hormones, unpredictable and struggling to differentiate between self and other (mother and foetus/mother and infant).

Current medical practice and scientific knowledge about pregnant bodies tries to manage this leaky body through a rational account. Social research has followed suit and has failed to fully examine women's 'embodied' experiences of leaky bodies in pregnancy and childbirth. The researchers aimed to fill this void.

Method and sample

To acquire an embodied account of pregnancy and labour, the researchers interviewed women about their experiences of their bodies.

The researchers longitudinally interviewed 25 women and their male partners in Sydney, Australia. Most couples (21 of them) were conveniently sampled, recruited at an antenatal class. Four further couples were recruited through the researchers' personal contacts. Participants were selected only if they were expecting their first child. Female respondents ranged from 23 to 35 years old. Most women were employed in white-collar occupations and had professional careers.

Semi-structured interviews were used. The couples were interviewed together, by one of the authors and another investigator. Nine interviews in total took place over three years, with an initial interview in late pregnancy and a second interview 2–10 days after birth. The researchers held the interviews close to the date of the infant's birth so that the women could describe their childbirth experience in detail (give a birth narrative) and describe how they felt about the way they gave birth.

> **Question**
>
> To what extent do you think 'embodiment' research can fully understand bodily experiences using methods like interviews? Explain your answer.

Findings

The women's birth narratives differed depending on the type of childbirth they experienced. Women who had given birth vaginally and without epidural (anaesthetic pain relief), and those who had a Caesarean section, gave the most detailed accounts. Six themes derived from the interview data are outlined below.

1. **Loss of control and the painful opening of the maternal body:** Women discussed the pain associated with childbirth and the awareness of the boundaries between the maternal and infant body, and the maternal body and the outside being more heightened:

   Marianne described the emergence of the placenta as follows: 'I remember that coming out, along with just about everything else, I felt like my insides fell out'.

   Jane discussed her sense of feeling her body 'open' to the world as its boundaries were stretched and widened, and the vulnerability this process engendered for her: The fact that you have to open up so much – you feel raw and exposed to yourself and to others and there is this expectation of just having to open up more. (pp. 831–832)

2. **Ambiguity and the unknown:** The process of childbirth was a blurring of boundaries between the mother and infant that could not be captured entirely by rational scientific and medical terminology:

   Technical terms do not fully encapsulate the labouring woman's physical and conceptual experiences that during this transitional phase, this body is liminal, neither fully one nor the other. (p. 832)

3. **Relief at expelling the 'other's' body:** Women who gave birth vaginally without anaesthetic expressed physical relief after the pain and discomfort of childbirth once the infant was born.

4. **The strangeness of the infant's body:** Instead of instant connections to their infants, many mothers felt the baby's body was strange:

   Tess: The midwife handed her straight to me and I held her, but I had held her for a while, I just was – it was like looking at her and wondering 'Where did this baby came from?' You know, despite what I'd gone through, it was hard to associate that she was actually mine and she was out of my stomach ... Even holding her for the first few minutes are just, it wasn't like she was mine, my kid, which is weird ... when you think of what you went through, it was really quite strange. (Tess)

Kerry: Oh, I was just overcome, like, 'Where did it [the baby] come from?' My support people both laughed at me later on because they said, 'You just, like it was as if like, wow, I didn't know that was going to come out!' I didn't know a baby was going to come out. It was just really spacey, a weird thing. (Kerry)

When the baby had come out and they had put it on me all bloody, I said 'Get him off me! Sorry', I said, 'Can you clean him up?' I just couldn't – with all that blood it was just so disgusting. And I thought maybe when it came out and it was all bloody I'd really want to hold it, but I didn't. (Amanda) (p. 834)

> **Questions**
>
> In what ways do you think embodied accounts of experiences such as this may differ to the data gathered through a traditional interview?
>
> What did this study reveal that Oakley did not capture in her study of motherhood in Chapter 5?

5. **The absent maternal body:** Women who underwent Caesarean sections often reported disembodiment – their bodies were anaesthetised, some under a local anaesthetic and some general anaesthetic. Even those under a local anaesthetic and aware of the operation were numbed and not fully aware of the childbirth process, to the extent that the process sometimes did not feel real.

6. **The absent infant body:** Caesarean sections also led women to report on the absent infant who was suddenly no longer in their body, or taken away from them after birth for checks.

The researchers suggested that their focus on embodied experiences provided insight into the process and feelings associated with childbirth that relate to what Kristeva (1982) called **abject bodies**, which are said to create feelings of unease and revulsion, and to inhabit a status 'in between' that disturbs identity and order.

Evaluation

The study gave insight into the 'embodied' experiences of women – that is, what it is like to be in their bodies at the time of pregnancy and childbirth. This provided new sociological insight into a topic that had typically been approached with less focus on the experience of childbirth. The sample size was relatively small and therefore arguably not representative. It is also interesting to reflect on whether the findings are peculiar to a Eurocentric perspective and experience.

While the interview method would score highly on validity, this might be improved if the women were interviewed without their partners. Due to the social embarrassment and stigma often associated with the 'leaky' aspects of childbirth, one might question whether the women would have been more open in their responses without their partner present.

CONTEMPORARY RESEARCH: EDUCATION: DECOLONISING METHODS TO REVEAL THE HIDDEN CURRICULUM

MacDonald, L. (2019). '"The same as everyone else": How academically successful indigenous secondary school students respond to a hidden curriculum of settler silencing'. *Whiteness and Education*, 4(1), pp. 38–52.

Study overview

This study focused on how the hidden curriculum constrains the Māori indigenous population of New Zealand. 'Hidden curriculum' refers to the informal learning that takes place in school and which tends to reinforce dominant social norms and values. Part of the hidden curriculum in this study was 'settler **silencing**', which refers to the exclusion of the meanings and effects of 'race' from state education to protect **White normativity** and White privilege (in this case, the Pākehā – the Māori term for the White dominant group). The aim of silencing is to ensure that **settler-colonial societies** continue to be ignorant about the historical violence they have enacted, as well as their ongoing racial domination. The study aimed to understand how indigenous students had agency and 'spoke back' to settler silencing.

Method and sample

Four semi-structured in-depth interviews were carried out with indigenous high-achieving Māori girls in one majority-White secondary school.

The concept of 'high-achieving' was measured by the attainment of Merit or Excellent in New Zealand national examinations during the prior academic year. The students were from middle-class backgrounds with supportive parents.

Interview transcripts were subject to **member-checking**, where the students looked over the transcripts and confirmed that the data was correct.

Kaupapa Māori methodologies guided the interviews. This is an approach that requires constant researcher reflections on how power shapes the different stages of the research process – for example, who is the research for? What difference will the research make?

Question

To what extent do you think the sample size would produce valid and reliable results?

Question

Why do you think it was important that the researcher used Kaupapa Māori methodologies to guide the interviews?

Key concepts

member-checking the process of having research participants check data throughout different stages of the research process to confirm accuracy and to enhance validity

Kaupapa Māori methodologies an approach to research that involves centring the question of the power relations at play during the research process; it requires the researcher to acknowledge the validity of Māori knowledge and incorporate it, along with a Māori worldview, into the interpretation and analysis of data

The core feature of this methodological approach is to look at the representation of indigenous people and communities throughout the research process.

The research method involved a thematic analysis underscored by critical race theory – a realist and constructionist epistemological position. Critical race theory centres the counter-narratives of ethnically and culturally marginalised groups to challenge the dominant culture's normalised worldview. It also requires the researcher to reflect on the ways that 'race' and power shape the research process itself. This includes the researcher recognising their own place in a racialised society. Blaisdell (2016) says this is important to avoid **dysconscious racism**, which is where the social context of the research study is assumed to be racially neutral.

Findings

MacDonald's analysis revealed three main ways in which the students responded to the hidden curriculum of settler silencing:

1. **Aligning with institutionally constructed ways of being Māori:** The students reported that the school organised cultural activities and that this gave them a foothold in a Māori identity. However, having a Māori identity within school was also problematic because it tended to 'fix' the girls' identities, framing them as 'other' against the dominant educational norms that equated White (Pākehā) with educational success and Māori with underachievement. They were also subject to other stereotyping within this fixing of identity in institutionally constructed ways of being Māori.

2. **Challenging institutional constructions of indigeneity:** The students were keen to be treated the same as everyone else and not have 'special treatment'. Where Māori students were offered extra help by the school, outside of formal lessons, and where the school engaged in culturally responsive teaching, the students found this treatment problematic as it constructed their identities as 'different'.

3. **Reproducing the whiteness of the institution to be academically successful:** The students hid aspects of their Māori identity to 'pass' in the educational institution, because they saw being Māori as an obstacle to educational success. Two of the students who were fair-skinned and could pass as White chose not to disclose, or tried to hide, their Māori identity at school. Because the students had multiple identities, they were able to select which ones to perform at school and which ones to perform elsewhere (Māori). They wore masks:

> *For the girls in this study, thinking or being Pākehā created pathways of resistance to negative stereotypes about their character or academic abilities which supported their drive for success. These pathways required that they assimilate Pākehā traits at school but assert the right to be Māori in other contexts. In this way, a naturalised hidden curriculum of settler silencing assures that 'whiteness is what the institution is orientated "around", so that even bodies that might not appear white still have to inhabit whiteness, if they are to get "in".' (p. 48)*

Evaluation

The research clearly addresses an important sociological issue – that of the colonial power structures that continue to shape education and discriminate against ethnic minority students. The research has validity because it uses in-depth interviews, but it has a very small sample size, which reduces its representativeness. Ethically, the research is highly reflexive of the various power relations at play throughout the entire research process, and the researcher was aware of her own positionality and the need to minimise and mitigate hierarchical power relations between herself and the students she researched.

Case study: Masks of performance in education

That children from minoritised ethnic groups are often framed as 'underachievers' has been revealed by other studies. Where MacDonald focused on how children put on masks to navigate the educational system, a study by Rollock et al. (2014), focused on parents. They undertook 80 interviews with Black Caribbean middle-class parents to find out how parents navigated their children's successes throughout the education system in Britain. The study showed how the Black British middle classes silenced some aspects of their working-class and cultural backgrounds in order to be accepted by White middle class teachers and parents. Various strategies were used to negotiate White educational spaces, including appealing to White sensibilities through language, 'code switching' or donning a 'mask' (p. 151).

Conclusion

This book has introduced you to a range of methods that we have referred to as a methodological kaleidoscope. Each turn of the kaleidoscope reveals new and important insights into the social world. Some researchers take just one turn of the kaleidoscope in each study, and may use just one method to reveal a different social pattern, while others take multiple turns, using mixed methods to reveal more of the 'picture' of social life.

This chapter has also introduced studies that use alternative, imaginative, creative and ethically imperative methods that continue to push social research forward, calling for social researchers to become more self-reflexive, more ethical and more socially responsible.

Social research methods (and the social researchers undertaking them) are socially situated, historical products that require critical attention and constant scrutiny if they are to stand the test of time and deliver not only rigorous data (representative, valid and reliable), but also data that is sociologically important, impactful and ethical. Feminist sociologists like Oakley (see Chapter 5) challenged methods that had previously centred social research around men. She revealed that much sociological research had been conducted through 'male ways of knowing' that privileged certain epistemological positions (positivism) and forms of knowledge historically associated with masculinity. Even Oakley's insight, which was seen to be trailblazing at the time of its publication, has been subject to further reflexive scrutiny. In this chapter, Lupton broadens feminist insights to include embodiment.

Sociologists today are still calling for further queering and decolonising of methods – finding methods that can be responsive to other ways of knowing, and that are not restricted to Eurocentric, anthropocentric, heteronormative and cisnormative forms of knowledge. These appeals for socially responsible and responsive research methods are core to the sociological project and mean that sociological research methods have never been more important.

CHAPTER 11: A FINAL TURN OF THE KALEIDOSCOPE: THE ENDURING IMPORTANCE OF SOCIAL RESEARCH

In 2007, Savage and Burrows argued that there is a 'coming crisis of empirical sociology'. In their words:

> Sociologists have not adequately thought about the challenges posed to their expertise by the proliferation of 'social' transactional data which are now routinely collected, processed and analysed by a wide variety of private and public institutions. [...] whereas over the past 40 years sociologists championed innovative methodological resources, notably the sample survey and the in-depth interviews, which reasonably allowed them to claim distinctive expertise to access the 'social' in powerful ways, such claims are now much less secure. [...] the sample survey and the in-depth interview are increasingly dated research methods, which are unlikely to provide a robust base for the jurisdiction of empirical sociologists in coming decades. (p. 885)

Savage gives a personal account of his realisation of the 'coming crisis' while at a research festival, talking to other participants interested in social network methods. After detailing his own research project, which he described as time-consuming and intensive, involving a postal questionnaire distributed to 320 members of three large organisations, alongside 30 in-depth life history interviews, he spoke to another researcher involved in social research studies:

> It turned out that one enthusiast was not an academic but worked in a research unit attached to a leading telecommunications company. When asked what data he used for his social network studies, he shyly replied that he had the entire records of every phone call made on his system over several years, amounting to several billion ties. This is data which dwarves anything that an academic social scientist could garner. Crucially, it was data that did not require a special effort to collect, but was the digital by-product of the routine operations of a large capitalist institution. It is also private data to which most academics have no access. (p. 887)

While sociologists once had a distinctive methodological toolbox, Savage and Burrows argue that there has been a wide-scale expansion of companies that can collect 'social' data using survey and interview methods routinely, on a much larger scale than sociologists can. They believe that this forces sociologists to ask what the purpose, role and value of social research is.

In this book we have introduced you to a range of social research methods, from the survey and in-depth interviews to the use of official statistics and other secondary data sources, ethnographic research, documentary methods and experiments. We have shown you how sociological knowledge is different from common-sense assumptions about the social world that non-sociologists also inhabit.

Savage and Burrows are correct that the social world is changing – there has been a marked rise in large datasets, as you saw in Chapter 3 (official statistics) and Chapter 7 (documentary methods) – and there is a wealth of material already available to sociologists. This data *does* exist in a world that non-sociologists inhabit, and some of it is even produced by non-sociologists. But does this mean that sociology is redundant? Is it true that social research methods, as they are employed by sociologists on often much smaller scales than non-sociologists, are unnecessary? Is sociological expertise unnecessary to conduct social research? We clearly do not think so!

It is true that the social survey is a declining method used by sociologists. We have noticed this ourselves in the writing of this book. Trying to find examples of sociological surveys that meet the STRIVE criteria is challenging. Today, far more sociologists use large, existing quantitative datasets to generate their analyses of all features of social life, ranging from families and households, to crime, education and inequality. Most of this data is already freely available to sociologists and is collected on a wide scale, through surveys that have been tested and are thus deemed reliable, valid and representative. It is therefore easy to see why sociologists would use them rather than generate their own datasets on a much smaller scale. However, this does not mean that social researchers are redundant. First, they continue to produce rich data and understanding that cannot be captured by pre-existing datasets. Second, they provide critical analysis and interpretation of existing datasets.

Sociological imaginations/social lenses

Each turn of a kaleidoscope gives a different view of the social world, but the picture also very much depends on who is looking. We argued at the start of this book that the first rule of STRIVE is that social research must be committed to being properly sociological. Suggesting that large datasets can replace sociologists reduces social methods to data processing – it quantifies the research process, regarding social research as purely procedural, efficient and effective. It is a positivist evaluation of methods, rather than a reflexive one.

As you saw in Chapter 1, we believe that research methods must meet the STRIVE criteria: they must be sociological, theoretical, rigorous, impactful, valid and reliable, and ethical and reflexive. Let's take each again in turn in response to Savage and Burrows.

Sociological

Large datasets may produce social data, but this data may not be sociological. That is, datasets have to be used correctly and interpreted with care. It is the questions that sociologists ask that highlight socially important relationships between variables within the data, which in turn sheds light on social inequalities. Sociologists turn descriptive social data into analytical and imaginative sociological understanding.

Theoretical

Social research is informed by different theoretical dispositions. It is reflexive and critical of its own orientation to research. Sociologists undertaking social research from a positivist orientation may be less inclined towards this reflexive disposition, but this does not mean that there is no need, or indeed imperative, for it. Savage and Burrows argue that sociologists cannot rely on their theoretical interpretations of data to differentiate themselves from non-sociologists. However, it is the sociologist's theoretical disposition that enables them to ask questions about the data – who produced it and whose interests sit behind it. More than this, social theory forces us to reflect on our own methods, beliefs, values and ways of doing and ways of seeing the world that would otherwise be blinkered. Theory differentiates sociological knowledge from common-sense assumptions.

Representative

Although large datasets generated by non-sociologists may encompass much larger samples than sociologists could ever hope to achieve on their own, this does not mean that sociologists are redundant, or that smaller samples are of little consequence. Small samples can generate rich and insightful data, and can be representative of the population or group under study. Likewise, new and innovative methodologies drawing increasingly on personal experiences add to the richness of sociological insight into social life. By focusing on big data alone, there is a danger that the thick description that qualitative methods afford is lost. To reduce our understanding of the social world to quantitative data would be a travesty.

Impactful

Some social databases are gathered by large organisations, including national governments, and are focused on impact – even more so than some social research undertaken by sociologists. However, we cannot be sure that the impact is not subject to institutional bias and that it is non-partisan in the same way that sociological research aims to be. Of course, social researchers have their own biases, but sociologists are reflexive in outlining these during the research process. Sociologists also see research impact as a constant process of reflection, rather than a final destination. Moreover, they ask questions that other people are not necessarily interested in, they unsettle accepted ways of thinking about the world and challenge the hegemonic power of people, authorities and governments.

Valid and reliable

While some social researchers use existing datasets, they are always mindful of challenges such as how concepts have been operationalised, whether these measures and indicators are socially robust, and whether there are any challenges with the measures themselves. Sociologists therefore hold social research to account – including the very measures of validity and reliability. Other methods, such as ethnographies, may be mimicked by 'mock ethnographic' documentaries, like those created by TV personalities such as Stacey Dooley, but unlike the social researcher, the documentary maker has not used systematic observation sheets or taken detailed thick descriptions in field notes. Nor have they necessarily followed ethical protocols, taken steps to 'member-check' data or used other triangulating methods. Social research, by contrast, takes all the necessary steps to ensure that data collection is as rigorous as possible.

Ethical and reflexive

Social research can be gathered and generated by large organisations, but many of these organisations are not subject to the same institutional ethical regulations as sociologists conducting social research. Such regulations ensure that ethical issues are minimised and that any possible harms to participants are always transparent and mitigated as fully as possible. Sociologists also reflect on and hold up to scrutiny the very methods and methodologies that social research uses – as you saw in Chapter 10, with regard to decolonial methodologies. As Matsinhe (2007) writes:

> *Methodology [...] legitimates and delegitimates, validates and invalidates, approves, and disapproves, passes and fails, claims to knowledge and knowledge production. Methodology is the final court of appeal in judging what counts as bona fide knowledge of something. (p. 838)*

Research methods themselves are constantly evolving, and social researchers must be aware of their histories – the assumptions underpinning them, the power relations sustaining them, and the inequalities that might be perpetuated by them, as well as the possibilities and promises that they bring.

The final turn

In our sister book, *How to Be a Sociologist*, we spoke of sociology's unique position as a discipline to be conceptual, rigorous, reflexive, knowledgeable, transformative, and – last but by no means least – hopeful. As the next generation of social researchers, we hope that you, too, see the enduring value and importance of sociology, and with this strive to support and celebrate the enduring value and importance of social research methods.

REFERENCES

Atkinson, J. M. (1973). 'Suicide, Status Integration and Pseudo-Science'. *Sociology*, 7(30), pp. 437–445.

Baker, R., Dee, T., Evans, B. and John, J. (2022). 'Bias in online classes: Evidence from a field experiment'. *Economics of Education Review*, 88, 102259. Available at: https://doi.org/10.1016/j.econedurev.2022.102259.

Ball, S., Maguire, M. and Braun, A. (2012). *How Schools Do Policy: Policy Enactments in Secondary Schools*. London: Routledge.

Barker, E. (1984). *The Making of a Moonie: Choice or Brainwashing?* Oxford: Blackwell.

Bauman, Z. (1990). *Thinking Sociologically*. Oxford: Blackwell.

Beauchez, J. (2021). 'Becoming a skinhead: An ethnobiography of brutalized life and reflective violence'. *The Sociological Review*, 69(6), pp. 1179–1194.

Beck, U. (2002). 'Zombie categories: An interview with Ulrich Beck'. In U. Beck and E. Beck-Gernsheim. *Individualization: Institutionalized Individualism and its Social and Political Consequences*. London: Sage, pp. 202–213.

Bejarano, C. A., Juárez, L. L., García, M. A. M. and Goldstein, D. M. (2019). *Decolonizing Ethnography: Undocumented Immigrants and New Directions in Social Science*. London: Duke University Press.

Berghs, M. (2021). 'Who Gets Cured? COVID-19 and Developing a Critical Medical Sociology and Anthropology of Cure'. *Frontiers in Sociology*, 5. Available at: https://doi.org/10.3389/fsoc.2020.613548.

Blaisdell, B. (2016). 'Exorcising the racism phantasm: Racial realism in educational research'. *The Urban Review*, 48, pp. 285–310.

Bloom, S. G. (2021). *Blue Eyes, Brown Eyes*. California: University of California Press.

Bourdieu, P. (1979). *Distinction: A Social Critique of the Judgement of Taste*. London: Routledge.

Bowling, B. and Phillips, C. (2007). 'Disproportionate and Discriminatory: Reviewing the Evidence on Police Stop and Search'. *Modern Law Review*, 70(6), pp. 936–961.

Bradley, H. (2014). 'Class Descriptors or Class Relations? Thoughts Towards a Critique of Savage et al.' *Sociology*, 48(3), pp. 429–436.

Bristow, J., Cant, S. and Chatterjee, A. (2020) *Generational Encounters with Higher Education: The academic-student relationship and the university experience*. Bristol: Bristol University Press.

Brown, C. (2015). *Educational Binds of Poverty. The lives of school children*. London: Routledge.

Burgess, A., Donovan, P. and Moore, S. (2009). 'Embodying Uncertainty?: Understanding Heightened Risk Perception of Drink "spiking"'. *British Journal of Criminology*, 49(6), pp. 848–862.

Calnan, M., Cant, S. and Gabe, J. (1993). *Going Private. Why People Pay for their Health Care*. Milton Keynes: Open University Press.

Cant, S. (2018). 'Hysteresis, social congestion and debt: Towards a sociology of mental health disorders in undergraduates'. *Social Theory & Health*, 16, pp. 311–325.

Cant, S. and Chatterjee, A. (2022). 'Powerful or disempowering knowledge? The teaching of sociology in English schools and colleges'. *Sociology*. Available at: https://doi.org/10.1177/00380385221107299.

Cant, S. and Hardes, J. (2021). *How to Be a Sociologist*. London: HarperCollins.

Cant, S., Savage, M. and Chatterjee, A. (2020). 'Popular but peripheral: The ambivalent status of sociology education in schools in England'. *Sociology*, 54(1), pp. 37–52.

Cant, S. and Sharma, U. (1998). 'Reflexivity, Ethnography and the Professions (Complementary Medicine) Watching You Watching Me Watching You (and Writing about Both of Us)'. *The Sociological Review*, 46(2), pp. 244–263.

Cant, S. and Sharma, U. (1999). *A New Medical Pluralism: Complementary Medicine, Doctors, Patients and the State*. London: Routledge.

Casey, E. and Littler, J. (2021). 'Mrs Hinch, the rise of the cleanfluencer and the neoliberal refashioning of housework: Scouring away the crisis?', *The Sociological Review*, 70(3), pp. 489–505.

Compton, D.L., Meadow, T. and Schilt, K. (Eds). (2018). *Other, Please Specify: Queer Methods in Sociology*. Oakland: University of California Press.

Demie, F. (2021). 'The Educational Achievement of Black African Children in England'. School Research and Statistics Unit Education and Learning, Lambeth Local Authority.

Denscombe, M. (2010). *The Good Research Guide for Small Scale Research Projects* (4th edition). Buckingham: Open University Press.

Denzin, N. K. (2000). 'The practices and politics of interpretation'. In N. K. Denzin and Y. S. Lincoln (Eds), *The Sage Handbook of Qualitative Research* (2nd edition), pp. 897–922. Thousand Oaks, CA: Sage.

Denzin, N. K. and Lincoln, Y. S. (2005). 'Introduction: The discipline and practice of qualitative research'. In N. K. Denzin and Y. S. Lincoln (Eds), *The Sage Handbook of Qualitative Research* (3rd edition), pp. 1–32. Thousand Oaks, CA: Sage.

Ding, X., Brazel, D. M. and Mills, M. C. (2022). 'Gender differences in sleep disruption during COVID-19: cross-sectional analyses from two UK nationally representative surveys'. *BMJ Open*, 12(4). Available at: https://bmjopen.bmj.com/content/12/4/e055792

Dlamini, N. J. (2021). 'Gender-Based Violence, Twin Pandemic to COVID-19'. *Critical Sociology*, 47(4–5), pp. 583–590. Available at: https://doi.org/10.1177/0896920520975465

Du Bois, W.E.B. (1899/1996). *The Philadelphia Negro*. Pennsylvania: University of Pennsylvania Press.

Durkheim, E. (1897/2005). *Suicide: A Study in Sociology*, translated by J. A. Spaulding and G. Simpson. London: Routledge.

Epp, C. R., Maynard-Moody, S. and Hader-Markel, D.P. (2014). *Pulled Over: How Police Stops Define Race and Citizenship*. Chicago: University of Chicago Press.

Feyerabend, P. (1975/2010). *Against Method* (4th edition). London: Verso.

Foucault, M. (1995). *Discipline and Punish: The Birth of the Prison*. New York: Vintage Books.

Francis, B. (2009). 'The Role of the Boffin as Abject other in Gendered Performances of School Achievement'. *The Sociological Review*, 57(4), pp. 654–669.

Francis, B., Craig, N., Hodgen, J., Taylor, B., Tereshchenko, A., Connolly, P. and Archer, L. (2020). The impact of tracking by attainment on pupil self-confidence over time: demonstrating the accumulative impact of self-fulfilling prophecy. *British Journal of Sociology of Education*, 41(5), pp. 626–642.

Friedman, S. and Laurison, D. (2020). *The Class Ceiling: Why it Pays to Be Privileged*. Bristol: Policy Press.

Garcia, R. and Tomlinson, J. (2020). 'Rethinking the Domestic Division of Labour: Exploring Change and Continuity in the Context of Redundancy'. *Sociology*, 55(2), pp. 300–318. Available at: https://doi.org/10.1177/0038038520947311.

Garfinkel, H. (1967/1991). *Studies in Ethnomethodology*. London: Wiley.

Garthwaite, K. (2016). *Hunger pains: Life inside foodbank Britain*. Bristol: Policy Press.

Geertz, C. (1973). *Thick Description: Toward an Interpretive Theory of Culture*. (1st edition). New York, NY: Basic Books.

Ghaziani, A. and Brim, M. (Eds). (2019). *Imagining Queer Methods*. New York: New York University Press.

Giddens, A. (1992/2013b). *The Consequences of Modernity*. London: John Wiley & Sons.

Gillborn, D., Demack, S., Rollock, N. and Warmington, P. (2017). 'Moving the goalposts: Education policy and 25 years of the Black/White achievement gap'. British Educational Research Journal, 43(5), pp. 848–874.

Goffman, A. (2014). *On the Run: Fugitive Life in an American City*. Chicago: University of Chicago Press.

Goldthorpe, J. and Hope, K. (1974). *The Social Grading of Occupations: A New Approach and Scale*. Oxford: Clarenden Press.

González, M. J., Cortina, C. and Rodríguez, J. (2019). 'The Role of Gender Stereotypes in Hiring: A Field Experiment'. *European Sociological Review*, 35(2), pp. 187–204. Available at: https://doi.org/10.1093/esr/jcy055.

Halberstam, J. (1998). 'An Introduction to Female Masculinity: Masculinity without Men'. In *Female Masculinity,* pp. 1–44. Durham, NC: Duke University Press.

Hall, S., Critcher, S., Jefferson, T., Clarke, J. and Roberts, B. (1978). *Policing the Crisis: Mugging, the State and Law & Order*. London: Macmillan Press.

Halls, A., Uprichard, E. and Jackson, C. (2018). 'Changing Girlhoods – Changing Girl Guiding'. *The Sociological Review*, 66(1), pp. 257–272.

Hanna, E. and Gough, B. (2020). 'The social construction of male infertility: a qualitative questionnaire study of men with a male factor infertility diagnosis'. *Sociology of Health & Illness*, 42(3), pp. 465–480. Available at: https://doi.org/10.1111/1467-9566.13038.

Haraway, D. (2006). 'A cyborg manifesto: Science, technology and socialist-feminism in the late twentieth-century'. In J. Weiss, J. Nolan, J. Hunsinger and P. Trifonas (Eds), *The International Handbook of Virtual Learning Environments*. Dordrecht: Springer.

Hardes, J. (2016). *Law, Immunization and the Right to Die*. London: Routledge.

Hardes, J. (2020). 'Governing excess: Boxing, biopolitics and the body'. *Theoretical Criminology*, 24(4), pp. 689–705.

Hochschild, A. R. and Machung, A. (1989). *The Second Shift: Working Parents and the Revolution at Home*. New York: Viking.

Humphreys, L. (1975). *Tearoom Trade: Impersonal Sex in Public Places*. New Jersey: Transaction Publishers.

Jennings, W. et al. (2021). 'Lack of Trust, Conspiracy Beliefs, and Social Media Use Predict COVID-19 Vaccine Hesitancy'. *Vaccines*, 9(6). p. 593. Available at: https://doi.org/10.3390/vaccines9060593.

Kan, M. Y. and Laurie, H. (2018). 'Who is Doing the Housework in Multicultural Britain?' *Sociology*, 52(1), pp. 55–74.

Keizer, K., Lindenberg, S. and Steg, L. (2008). 'The Spreading of Disorder'. *Science*, 322(5908), pp. 1681–1685.

Keuschnigg, M. and Wolbring, T. (2015). 'Disorder, social capital, and norm violation: Three field experiments on the broken windows thesis'. *Rationality and Society*, 27(1), pp. 96–126.

Khattab, N. and Modood, T. (2017) 'Accounting for British Muslim's educational attainment: Gender differences and the impact of expectations', *British Journal of Sociology of Education*, 39(2), pp. 242–259.

Koch, I., Fransham, M., Cant, S., Ebrey, J., Glucksberg, L. and Savage, M. (2020). 'Social Polarisation at the Local Level: A Four-Town Comparative Study on the Challenges of Politicising Inequality in Britain' *Sociology*, 55(1), pp. 3–29. Available at: https://doi.org/10.1177/0038038520975593.

Kuhn, T. (1970). *The Structure of Scientific Revolutions*. Chicago: University of Chicago Press.

Lakatos, I. (1978/2001). *The Methodology of Scientific Research Programmes*. Cambridge: Cambridge University Press.

Lau-Layton, C. (2014). *British Chinese Families: Parenting Approaches, Household Relationships and Childhood Experiences*. London: Palgrave Macmillan.

Lincoln, Y. S. (1995). 'Emerging criteria for quality in qualitative and interpretive research'. *Qualitative Inquiry* (1), pp. 275–289.

Lincoln, Y. S. and Guba, E. G. (1985). *Naturalistic Inquiry*. Beverly Hills, CA: Sage.

Lupton, D. (2019). *Data Selves: More-than-Human Perspectives*. Cambridge: Polity Press.

Lovo, E., Woodward, L., Larkins, S. et al. 'Indigenous knowledge around the ethics of human research from the Oceania region: A scoping literature review'. *Philos Ethics Humanit Med* 16, 12 (2021). Available at: https://doi.org/10.1186/s13010-021-00108-8

Lupton, D. and Schmeid, V. (2013). 'Splitting bodies/selves: Women's concepts of embodiment at the moment of birth'. *Sociology of Health & Illness,* 35(6), pp. 828–841.

Lupton, D. and Watson, A. (2021). 'Towards more-than-human digital data studies: developing research-creation methods'. *Qualitative Research*, 21(4), pp. 463–480. Available at https://doi.org/10.1177/14687941209392.

MacDonald, L. (2019). '"The same as everyone else": How academically successful indigenous secondary school students respond to a hidden curriculum of settler silencing'. *Whiteness and Education*, 4(1), pp. 38–52.

Maestripieri, L. (2021). 'The COVID-19 Pandemics: Why Intersectionality Matters'. *Frontiers in Sociology*, 6. Available at: https://doi.org/10.3389/fsoc.2021.642662.

Malinowski, B. (1929/2012). *The Sexual Life of Savages in North-Western Melanesia: An Ethnographic Account of Courtship, Marriage, and Family Life Among the Natives of the Trobriand Islands, British New Guinea*. London: Routledge.

Matsinhe, D. (2007). 'Quest for Methodological Alternatives'. *Current Sociology*, 55(6), pp. 836–856.

Matthewman, S. and Huppatz, K. (2020). 'A sociology of Covid-19'. *Journal of Sociology*, 56(4), pp. 675–683. Available at: https://doi.org/10.1177/1440783320939416.

McArthur, D. and Reeves, A. (2019). 'The Rhetoric of Recessions: How British Newspapers Talk about the Poor When Unemployment Rises, 1896–2000'. *Sociology*, 56(5), pp. 1005–1025.

McKnight, A. (2015). *Downward mobility, opportunity hoarding and the 'glass floor'*. Research report. Social Mobility and Child Poverty Commission.

Meghji, A. (2021). *Decolonizing Sociology: An Introduction*. Cambridge: Polity Press.

Mijs, J. and Savage, M. (2020). 'Meritocracy, elitism and inequality'. *The Political Quarterly*, 91(2). pp. 397–404.

Miller, T. (2011). 'Falling back into gender? Men's narratives and practices around first-time fatherhood'. *Sociology*, 45(6), pp. 1094–1109.

Mills, C. W. (1959/2000). *The Sociological Imagination*. Oxford: Oxford University Press.

Oakley, A. (1974/2019). *The Sociology of Housework*. Bristol: Policy Press.

Oakley, A. (1979/2019). *From Here to Maternity. Becoming a Mother*. Bristol: Policy Press.

Oakley, A. (2016). 'Interviewing Women Again: Power, Time and the Gift'. *Sociology*, 50(1), pp. 195–213.

Oates, C. J. and McDonald, S. (2006). 'Recycling and the Domestic Division of Labour: Is Green Pink or Blue?'. *Sociology*, 40(3), pp. 417–433.

Office for National Statistics (2018). *National Measurement of Loneliness: 2018*. https://www.ons.gov.uk/peoplepopulationandcommunity/wellbeing/compendium/nationalmeasurementofloneliness/2018.

Orgad, S. (2019). *Heading Home: Motherhood, Work, and the Failed Promise of Equality*. New York: Columbia University Press.

Pande. R. (2021). *Learning to Love: Arranged Marriages and the British Indian Diaspora*. New Jersey: Rutgers University Press.

Parry, M. (2019). 'Alice Goffman's first book made her a star. It wasn't enough to get her tenure' [online]. Available at: https://www.chronicle.com/article/alice-goff mans-first-book-madeher-a-star-it-wasnt-enough-to-get-her-tenure/.

Phillips, C. (2019). 'The trouble with culture: A speculative account of the role of Gypsy/Traveller cultures in "doorstep fraud"'. *Theoretical Criminology*, 23(3), pp. 333–354.

Plummer, K. (2005). 'Critical humanism and queer theory'. *The Sage Handbook of Qualitative Research*, 3, pp. 357–373.

Plummer, K. (2013). 'A manifesto for a critical humanism in Sociology'. In D. Nehring (Ed.), *Sociology: A Text and Reader*. London: Pearson.

Poole, E. and Williamson, M. (2021). 'Disrupting or reconfiguring racist narratives about Muslims? The representation of British Muslims during the Covid crisis'. *Journalism*, pp. 1–18. Available at: https://doi.org/10.1177/14648849211030129.

Popper, K. (1935/2002). *The Logic of Scientific Discovery*. Oxon: Routledge.

Quadlin, N., Jeon, N., Doan, L. and Powell, B. (2022). 'Untangling perceptions of atypical parents'. *Journal of Marriage and Family*, 84(4), pp. 1175–1195.

Richardson, L. (2000). 'Writing: A method of inquiry'. In N. K. Denzin and Y. S. Lincoln (Eds), *The Sage Handbook of Qualitative Research* (2nd edition), pp. 923–48. Thousand Oaks, CA: Sage.

Rollock, N., Gillborn, D., Vincent, C. and Ball, S. (2014). *The Colour of Class: The Educational Strategies of the Black Middle Classes*. New York: Routledge.

Rosenthal, R. and Jacobson, L. (1968). 'Pygmalion in the classroom'. *The Urban Review*, 3(1), pp. 16–20.

Savage, M. (2015). *Social Class in the 21st Century*. London: Penguin.

Savage, M. and Burrows, R. (2007). 'The Coming Crisis of Empirical Sociology'. *Sociology*, 41(5), pp. 885–899.

Scambler, G. (2020). 'Covid-19 as a "breaching experiment": exposing the fractured society'. *Health Sociology Review*, 29(2), pp. 140–148.

Scott, J. (1990). *A Matter of Record: Documentary Sources in Social Research*. Cambridge: Polity Press.

Shain, F. (2011). *The New Folk Devils: Muslim Boys and Education in England*. Stoke-on-Trent: Trentham Books.

Shain, F. (2021). 'Navigating the unequal education space in post-9/11 England: British Muslim girls talk about their educational aspirations and future expectations'. *Educational Philosophy and Theory*, 53(3), pp. 270–287.

Shildrick, M. (1997). *Leaky Bodies and Boundaries: Feminism, Postmodernism and (Bio)ethics*. London: Routledge.

Shilling, C. (2012). *The Body and Social Theory* (Third Edition). London: Sage.

Sinclair, I. (2014). 'Forty years on from the publication of her landmark book Housewife, Ann Oakley talks to the Morning Star modern feminism and gender roles today'. *Morning Star Online*, accessed at: https://morningstaronline.co.uk/a-af88-the-changing-role-of-the-housewife-1.

Singh, G. (2023). 'Sikh ethics: A guide for professionals and researchers'. Available at: https://asiasamachar.com/2023/02/01/sikh-ethics-a-guide-for-professionals-and-researchers. Accessed 3/5/23

Singha, L. (2019). *Work, Labour and Cleaning: The Social Contexts of Outsourcing Housework*. Bristol: Policy Press.

Smart, C. (2011). 'Ways of Knowing: crossing species boundaries'. *Methodological Innovations Online*, 6(3), pp. 27–38.

Smith, L. T. (2013). *Decolonizing Methodologies: Research and Indigenous Peoples* (3rd edition). London: Zed Books Ltd.

Stacey, L. (2021). '(Bio)logics of the Family: Gender, Biological Relatedness, and Attitudes Toward Children's Gender Nonconformity in a Vignette Experiment'. *Sociological Forum*, 37(1), pp. 222–245.

Strand, S. (2012). 'The White British-Black Caribbean Achievement Gap: Tests, Tiers and Teacher Expectations'. *British Educational Research Journal,* 38(1), pp. 75–101. Available at: https://doi.org/10.1080/01411926.2010.526702.

Strand, S. (2015). 'Ethnicity, deprivation and educational attainment gap at age 16 in England: trends over time'. London: Department for Education.

Szabo, M. (2013). 'Foodwork or Foodplay? Men's Domestic Cooking, Privilege and Leisure'. *Sociology*, 47(4), pp. 623–638.

Taylor, R. (2009). 'Slain and slandered: A content analysis of the portrayal of femicide in crime news'. *Homicide Studies*, 13(1), pp. 21–49.

Thebauld, S., Kornrich, S. and Ruppanner, L. (2021). 'Good Housekeeping, Great Expectations: Gender and Housework Norms'. *Sociological Methods & Research*, 50(3), pp. 1186–1214. Available at: https://doi.org/10.1177/0049124119852395.

Thomas, G., Lupton, D. and Pedersen, S. (2018). '"The appy for a happy pappy": Expectant fatherhood and pregnancy apps'. *Journal of Gender Studies*, 27(7), pp. 759–770.

Townsend, P. (1979). *Poverty in the United Kingdom*. London: Allen Lane and Penguin Books.

Ujomudike, P.O. (2016). 'Ubuntu Ethics'. In: ten Have, H. (eds) *Encyclopedia of Global Bioethics*. Springer, Cham. https://doi.org/10.1007/978-3-319-09483-0_428

Venkatesh, S. (2008). *Gang Leader for a Day*. London: Penguin Press.

Venn, S., Arber, S., Meadows, R. and Hislop, J. (2008). 'The fourth shift: exploring the gendered nature of sleep disruption among couples with children'. *The British Journal of Sociology*, 59(1), pp. 79–97.

Waddington, P. A. J., Stenson, K. and Don, D. (2004). 'In Proportion: Race, and Police Stop and Search'. *British Journal of Criminology*, 44(6), pp. 889–914.

Walby, S., Towers, J. and Francis, F. (2014). 'The decline in the rate of domestic violence has stopped: Removing the cap on repeat victimisation reveals more violence'. Violence & Society Research Briefing. Lancaster University. Available at: https://eprints.lancs.ac.uk/id/eprint/72272/4/Violence_Society_Research_briefing_1.pdf. Accessed on 8/8/22.

Walter, M. and Andersen, C. (2013). *Indigenous Statistics: A Quantitative Research Methodology*. London: Routledge.

Willis, P. (1978/2016). *Learning to Labour: How working class kids get working class jobs*. London: Routledge.

Wood, M., Hales, J., Purdon, S., Sejersen, T. and Hayllar, O. (2009). 'A test for racial discrimination in recruitment practice in British cities'. Research report 607. London: Department for Work and Pensions.

Yavorsky, J. E., Qian, Y. and Sargent, A.C. (2021). 'The gendered pandemic: The implications of COVID-19 for work and family'. *Sociology Compass,* 15(6).

Young, J. (1999). *The Exclusive Society: Social Exclusion, Crime and Difference in Late Modernity*. London: Sage.

Zimbardo, P. G. (1973). 'On the ethics of intervention in human psychological research: With special reference to the Stanford prison experiment'. *Cognition*, 2(2), pp. 243–256.

GLOSSARY

abject bodies bodies that create feelings of unease, disgust and revulsion in others; social norms define what is experienced as abject, e.g. some people might experience revulsion at dead bodies, human excrement, the skin on milk, or dirt and grime

absolute poverty when a household income is not enough to sustain a basic standard of living and meet essential human needs

agency the notion that individuals have the capacity to act freely; often contrasted with determinism and social structure

alienation in Marxist theory, a process in which someone becomes disconnected from their work and from other humans

anthropocentrism the belief that human beings are the single most important entities in the world, and conceiving of everything in terms of humans

assemblage a collection or gathering of things, people and animals; assemblages are hybrids – nothing exists in isolation but always in relation with and to other things and people

bias a tendency to prefer one person or thing to another, and to favour that person or thing

big data information that is generated by companies such as Google; this data is vast and complex, and cannot be analysed using traditional sociological methods

binaries the division of the social world into two categories; these are often problematic as they simplify reality and reinforce dominant ideas, privileging one side of the binary over the other, e.g. men over women, straight over gay

black box a phenomenon that is taken for granted, so there is no sociological research into it; 'opening the black box' means researching something that has not been studied before

bourgeoisie the social class that owns the means of production and so is made up of more powerful members of society than other classes

broadsheet a newspaper that contains major news stories of higher quality than a tabloid newspaper; they provide more in-depth analysis, use more complex language and have lower circulation, e.g. *The Times*, *The Guardian*, *The Telegraph*

capital a set of skills and resources that take numerous forms; Bourdieu identifies five types of capital – cultural, economic, physical, symbolic, and social – and it is the accumulation of different types of capital that constitutes someone's social status

capitalism a system defined by its economic relations of production, where goods and services are bought and sold for profit, and in which two distinct classes emerge – those who own the goods and services (the bourgeoisie) and those who have to sell their labour to earn a living (the proletariat)

cisnormativity the dominant social belief that a person's gender corresponds with their biological sex at birth

closed society a society in which social mobility is rare or impossible

colonialism the process through which a country or group of countries takes over the land and people of another country or countries

consumerism an ideology of mass production that leads people to believe that consuming goods is essential to their identity

corporal punishment a physical punishment, such as hitting someone

criminalisation the process through which behaviours and individuals are transformed into crimes and criminals

critical disability studies studies that focus on challenging the socially constructed ableist assumptions that shape society and thus stigmatise and exclude 'disabled' bodies

critical race theory an academic movement made up of scholars and activists who examine how White privilege, social constructions of 'race' and racism continue to underscore inequity and injustice

cultural capital knowledge, lifestyle, choices, values, leisure activities and education

cultural determinism the belief that our culture determines how we behave; this is contrasted with biological determinism, which suggests that our inherited traits predict life chances and behaviour

cultural hegemony the dominance of the upper classes in a culturally diverse society

data selves derived from Lupton (2019), the process by which our understanding of ourselves becomes intertwined with the data that is held about us

diaspora people who come from a particular nation, or whose ancestors came from it, but who now live in many different parts of the world

discourse analysis the contextual examination and deconstruction of meanings conveyed in texts and verbal exchanges

discourses dominant knowledges that circulate about the social world, e.g. science, medicine, law

domestic division of labour the way that household tasks are divided, typically by gender but also by other social demographics, including ethnicity, ability, age and sexuality

dominant discourses knowledges that circulate about the social world, e.g. science, medicine, law

dualism a philosophical position that holds that there are two basic opposed principles, such as mind/body, structure/agency, man/woman, human/non-human; dualisms rely on sustaining hierarchies (one category is held as superior)

dysconscious racism a form of racism that tacitly accepts White normativity and White privilege

ecological validity the extent to which experimental findings can be generalised to the real world

economic capital how much income and wealth someone has

embodied describing the view that the body is not an object separate from the mind, but that mind and body are joined together and that we 'know' about the world through our bodies, not by distancing ourselves from them

ethical relating to beliefs about right and wrong

fat studies studies that focus on 'fatness' and the 'fat body' as a cultural construction and stigmatised identity

femininity characteristics typically associated with women and girls; feminists argue that these characteristics are shaped by and reinforce a patriarchal social order

feminist perspectives various theories that attempt to understand women's place in society (e.g. Black feminism, decolonial feminism, liberal feminism, Marxist feminism, patriarchy, radical feminism)

feudalism a social, economic and political structure that characterised the medieval period, where landowners held power over labourers (serfs), who worked for the lords or rented the land

functionalism a consensus theory which suggests that society requires social norms and customs to function; different social institutions (the family, Church, education, etc.) all have different functions in socialising people into the norms and customs of wider society, and they are interdependent – a dysfunction in one area of society can impact other areas

generalisable applicable to the wider population

Global North one division of the globe that acknowledges the social and economic differences that characterise nations; northern countries are more likely to be wealthy and powerful, in part due to the legacy of colonial exploitation

habitus the deeply ingrained habits, dispositions, norms, skills and other characteristics of a person that are defined by different types of capital (social, economic, cultural, etc.)

hegemonic describing the dominance of one social group (usually a minority) over another (usually the majority); described from a neo-Marxist perspective as a type of power exercised through ideologies that use language and persuasion to make the masses believe a certain 'truth', established by a ruling class, that conceals reality

hegemonic masculinity the legitimation of men's dominance over women through the use of gendered ideologies about how men should conform to gendered norms of masculinity and women should conform to gendered notions of femininity

heterogeneous describing a group that consists of many different types of people or things

heteronormativity the normalisation of heterosexuality and its associated character traits for both men and women

hidden curriculum the things students learn in school that are not part of the formal classroom content

highbrow culture intellectual taste that is suggestive of superiority, e.g. showing interest in serious art or complicated subjects such as opera or ballet

homogeneous describing a group whose members are all the same

indexicality the fact of language and other forms of expression having meaning only in relation to the context in which they are being used

individualisation the process through which individuals come to take responsibility for their own independent lives, at the expense of wider welfare systems and communal frames of reference; individualisation is said to have intensified in the era of neoliberalism

insider status when a researcher has similar socio-demographics, cultural values and experiences to the people they are researching

institutional racism a form of racism that is hard to see, but which occurs in the rules, decisions and practices of organisations such as schools, police forces, banks and health services

labelling applying terms to describe or categorise people that impact behaviour and sense of self (e.g. labelling someone as a 'naughty child' will reinforce naughty behaviour)

left wing a political position focused on advocating for social welfare support, equality, eliminating social inequality, and supporting social justice for minoritised populations

leisure gap the 'gap' experienced between men and women in terms of the amount of leisure (free) time they have outside paid employment and domestic work

LGBTQIA+ an acronym for lesbian, gay, bisexual, transgender, questioning, queer, intersex, asexual, pansexual and allies, although it is intended to reflect gender and sexual fluidity and thereby to be inclusive of people of all genders and sexualities

low-brow culture tastes that are not highly intellectual or cultured and which do not demand much intelligence to be understood, such as reality TV shows, bingo, pop music

macro describing things that occur on a large scale, typically referring to sociological theories that look at how large institutions and norms shape human conduct

marketisation a trend from the 1980s where schools (and other institutions) were encouraged to act like businesses and compete with one another

matrix of domination a concept coined by Patricia Hill Collins to explain the interconnections between oppressions relating to ethnicity, class and gender

materialism the theory that the material world that we can perceive with our senses has an objective reality independent of us

medicalisation the process through which human conditions and problems come to be defined as medical conditions

meritocracy the belief that achievements in life are based on ability and the effort that individuals put into things

micro describing sociological studies that examine the individual interactions between people

middle-brow culture moderately but not highly cultivated tastes and interests, such as musicals and national art galleries

monism a philosophical position that denies the existence of dualisms

moral panic widespread anxiety among the general public, usually relating to a certain condition or a group of people seen to threaten social norms; the media plays a significant role in generating moral panics by sensationalising and exaggerating an issue

narrative turn a 'turn' in the social sciences towards more story-like reporting of research data, in order to break with scientific 'detached' writing conventions

neoliberalism a political approach that promotes competition, free trade and a reduction in government spending, and places emphasis on each individual to be responsible for themselves

neologism a newly coined word or phrase

normative assumptions shared values and beliefs that have a moral dimension

open society a society in which people can change their social position and achieve higher social status and rewards when moving up the social ladder

opportunity hoarding when privileged groups control access to resources, such as places in grammar or private schools, and put practices in place to ensure they can take advantage of opportunities

othering the practice of socially excluding a group, creating a binary between 'us' and 'them'; the label 'other' defines someone as belonging to a subordinate category

paradigm 'world view' – the idea that over time different world views replace one another as ideas become scientifically accepted and more dominant; Kuhn referred to this process as 'scientific revolutions'

patriarchy the hierarchical organisation of society in which men are the dominant group and exercise power over women, who are the subordinate group

phenomenological describing a philosophical position that focuses on humans' lived experiences

physical capital bodily dispositions and shape (including how attractive someone is deemed to be)

post-humanism a theoretical position focused on breaking the boundaries between humans and non-humans, including animals and technology

post-structuralism sometimes used interchangeably with postmodernism, this is a philosophical approach that challenges fixed structures, truths, knowledges and realities, and instead emphasises revealing which truths and knowledges gain precedence in society and why

postcolonial describing a position that challenges historical narratives and traditions emerging from Europe that have historically seen and framed the world (and knowledge) through a colonial lens

postmodernism a theory that suggests humans have moved beyond modernity and are living in an entirely different world, where scientific standards of research should be interrogated, critiqued and challenged; it emphasises the idea of multiple realities and multiple truths

proletariat the social class of workers, who do not have their own means of production

quare studies research that foregrounds the complex intersections between sexuality and ethnicity

queer theory a perspective that explores and challenges the way that sex-based binaries between men and women dominate our thinking and social norms and are oppressive for those who cannot, or do not wish to, live according to such norms

race a social construct that historically emerged to categorise people according to physical features, such as skin colour, but which has no foundation in real biological differences between people

racism discrimination against people based on the socially constructed categories of 'race'

relative poverty when a household income is below the median income (midpoint) of a society

right wing a political position underscored by traditionalism, free will and private ownership

risk discourse (Beck) a preoccupation with the identification of risks, hazards and insecurities associated with living in the modern world

second shift the domestic responsibilities that women undertake alongside paid work; sometimes referred to as a 'double shift' or 'dual burden'

self-fulfilling prophecy when expectations about someone impact their behaviour and become true

settler-colonial society a distinct form of colonialism, in which indigenous society is replaced with a settler society that develops sovereignty, e.g. Canada, Australia, South Africa, New Zealand and the USA

silencing when a dominant group marginalises or excludes the voices of oppressed social groups

social capital a person's social networks, relationships and who they know

social facts the values, norms and social structures that shape people's behaviour and life chances (e.g. language, religion, marriage, laws and family)

social mobility the movement of people between social classes, either upwards or downwards (see also *meritocracy*, *capital*)

socialisation the process through which people learn social customs and norms – the acceptable and unacceptable ways of behaving in society

standpoint the view that all knowledge emanates from a social position

stigma a socially discredited attribute, behaviour or reputation that prevents someone from being socially accepted

subculture a grouping of people based on shared cultural ideas and identities

surveillance society a society in which individual or public behaviour is monitored and/or recorded; Foucault observes how these practices have an impact on our conduct and choices

symbolic capital a person's status and prestige

tabloid a newspaper that is image led, with shorter stories written in more simplistic language than a broadsheet; tabloids report on celebrity gossip as well as global news; 'red top' tabloids include *The Sun*, *Daily Mirror* and *Daily Star*; middle market tabloids include *The Daily Mail* and *Daily Express*

third shift the emotional work that women sometimes do in addition to career work and childcare; sometimes known as the 'triple shift'

transgender describing someone who has a different gender identity than the sex assigned at birth

unacceptable femininity characteristics, attributes or behaviours that conflict with or challenge feminine stereotypes

White normativity when whiteness is taken as the standard of the norm

white-collar crime non-violent crime, often financial, such as fraud or money laundering, which tends to be committed by members of the middle classes and elite

xenophobia a dislike or prejudice against people from other countries

zombie category a concept that is still widely used in sociology but lacks credibility or validity

INDEX

A
abject bodies 212
absolute poverty 76
achievement gap 58
animal-human relations 202–3
anonymity 30, 130, 147
anthropocentrism 201
a priori assumptions 97
Arber, S. 191–2
arranged marriages, ethnographic study of 136–9
attitudinal questions 68
atypical parents, public perceptions of 177–81
authenticity 151
authoethnography (ethnobiography) 9, 201, 207–9
auxiliary hypotheses 22

B
Ball, S. 160–1
Baker, R. 173–4
Barker, E. 189–90
Beauchez, J. 207–9
biases 6, 31
    in employment 174
big data 9, 14, 42, 146, 219
blog posts 146
Bloom, S.G. 172
body
    language 103
    decoration 208–9
    methods 202, 203
    women's (childbirth) 210–12
bodily habitus 202
Boffins 196–8
Bourdieu, P. 28, 78–80, 81, 85, 202, 208
bourgeoisie 18
Bowling, B. 60–2
breaching experiment 170–1
Braun, A. 160–1
Bristow, J. 148–9
British Crime Survey 62
British Muslims,
    education, 59, 112–14
    press representations during COVID-19 pandemic 150–1
British Sociological Association 30, 144, 147
broken windows theory (BWT) 182–3
Brown, C. 140–1
Burgess, A. 193–5
Burrows, R. 217–18

C
canny knowledge 203
Calnan, M. 188
Cant, S. 34, 45–6, 73, 101–3, 127, 148–9, 188
capital (economic, social, physical, cultural and symbolic) 57, 59, 68, 78–82, 83, 85, 95, 182–3
capitalism 6, 17, 147, 155, 199
Casey, E. 147
causation 16
Chatterjee, A. 148–9
childbirth, women's experiences of 105–7, 210–12
cisnormativity 11, 201, 217
class ceiling 50–4
cleanfluencer 147
closed questions 8, 69–70
closed society 53
Collins, P.H. 19
colonialism 10, 19, 31, 32, 71, 136, 139, 213–15
coming crisis of empirical sociology 14, 217
confidentiality 30, 130, 147, 209
conformability 29
constructionism (social) 15, 16, 17, 20–1, 29, 35, 36, 40, 44, 47, 49, 62, 93–4, 102, 111, 118, 122, 152, 156–7, 169, 187, 197, 201, 206, 214
constructionists 17
    ontological position 15
consumerism 148
content analysis (quantitative): 20, 148
convenience sampling 25–6
corporal punishment 119
correlations 13, 16, 39, 45, 46, 53, 72, 140, 169, 173, 174, 184
COVID-19 pandemic 4
    sociology of 5
credibility 29, 151
Crenshaw, K. 19
crime
    broken windows theory (BWT) 182–3
    gangs 132–4
    discourses 153
    domestic violence 63–4

doorstep 110–11
drink spiking 193–5
ethnicity and 60–2, 110–11, 132–4
femicide 158–9
media 154–5, 158–9
stop and search 60, 90–1
surveys 61–2, 63
understanding 153–5
white collar 18
criminalisation 111
critical disability studies 205
critical discourse analysis 152
critical heterosexualities 205
critical race theory 17, 19, 22, 58, 213
cultural deprivation 140
cultural determinism 111
cultural hegemony 80
cyborgs 202, 204

D

data selves (Lupton) 42, 204
date rape 193
deception 30, 130
decolonial methods 9, 136, 205, 220
    to reveal hidden curriculum 213–15
decolonising ethics 30–1
deduction 97
Dee, T. 173–4
dependability 29
descriptive questions 35
detachment 126
digital ethics 147
discourse analysis (qualitative) 9, 19–20, 101, 148
discourses 20
Doan, L. 177–81
documentary methods 20, 146, 170
    advantages of 165
    constructionist approach 152
    content analysis (quantitative) 148
    disadvantages of 165
    discourse analysis (qualitative) 148
    interpretivist approach 152
    positivist approach 152
    realist approach 152
    reliability and validity 151–2
    sets of criteria for assessing documents 151–2
    as a social research method 165
domestic division of labour 44, 192

domestic violence 63–4
dominant discourses 20
Donovan, P. 193–5
Dooley, S. 13
double-barrelled questions 71
double hermeneutic 32
drink 'spiking' 193–5
drug-facilitated sexual assault (DFSA) 193–5
dualism 202
Du Bois, W.E.B. 19, 74–5
Durkheim, É. 15–16, 21
    study of suicide 48–9
dysconscious racism 214

E

education
    boffins 196-8
    educational experiences of British Muslim girls and boys 8
    educational policy 160–1
    ethnic and gender bias in online classes 172–4
    ethnicity and education 8, 58–9, 112–14, 215
    hidden curriculum 213–15
    impact of identity on educational achievement 196–8
    masks of performance in 215
    meritocracy 50–7
    self-fulfilling prophecy and 86–89, 111, 169
    setting 86–7
    stereotyping of high-achieving students 196–8
educational outcomes
    correlation between parental social class and 15–16
    correlation between poverty and 140–1
Elliott, J. 9, 172
embodied methods 9, 202
embodiment 210–12
empirical data 17–18
epistemology 14–20
Epp, C. R. 90–1
ethics 13, 30, 38, 126
ethnicity
    Black British middle classes 215
    British Chinese family life 118–19
    class and 215
    crime and 60–2
    education and ethnicity 8, 58–9, 112–14,
    housework and 43–4
    intersectionality 19

media representation of 150–1, 154–5
occupation 50
ethnobiographic research 9, 201, 207–9
ethnographies 7, 124–5, 133
ethnomethodology 17, 170
evaluative questions 35
Evans, B. 173–4
experimental methods 166
    advantages of 184
    breaching experiment 170–1
    constructionist approach 169
    disadvantages of 184
    experimental design in surveys 1, 169
    field experiments 168
    interpretivist approach 169
    laboratory 168
    positivist approach 169
    realist approach 169
    as a social research method 184
    Stanford Prison Experiments 167
    Tuskegee Experiment 167
    validity and reliability 169
explanatory questions 35

F

facts (social) 16
falsification 21
families
    arranged marriage 136–9
    atypical 177–81
    attitudes to gender non-conformity 175–6
    fatherhood
        expectant, use of pregnancy apps 156–7
        first-time 108–9
    motherhood
        leaving paid employment, influencing factors 116–17
        long hours and competing demands of work 117
        work-life balance 117
        stay at home mothers 115–17
fat studies 205
female 'ways of knowing' 18
femicide portrayal, in crime news 158–9
feminism 22–3
    intersectionality and 19
feminist methods 10, 103–8
feminist realist epistemology 18
feudalism 18

Feyerabend, P. 23
field notes 125
first-time fatherhood 108–9
focus groups 100
foodbanks 142–4
Foucauldian Discourse Analysis 152
Foucault, M. 20, 153–4
Friedman, S. 50–54
Francis, B. 86–9, 196–9
Francis, F. 63–4
functionalism 15

G

Gabe, J. 188
Garcia, R. 97
Garfinkel, H. 31, 170–1
    breaching experiments 9
Garthwaite, K. 142–4
gatekeeper 25, 125
gay subculture 128–31
gender bias experiment 166
gender non-conformity 175–6
generalisable research 25
geneticist 6
Gilborn, D. 58
Girl Guides 164–5
girlhood 164–5
glass ceiling 50, 57
glass floor 54–7
Global North 19, 31
Global Southern knowledge 205
going native (over-immersion) 32
Goffman, A. 134
Gough, B. 92–4
Great British Class Survey (GBCS) 8, 81–5
grounded theory 17, 97
group discussions 100
Gypsy/Traveller cultures and association with doorstep fraud 110–12

H

habitus 78
Hader-Markel, D.P. 90–1
Halberstam, J. 206
Halls, A. 164–5
Hanna, E. 92–4
Hardes, J. 34, 149
harm 130

harmonised questions 69
Hawthorne Effect 32, 166
hegemonic masculinity 93, 94, 157, 198
heterogeneous examination 59
heteronormativity 11
hidden curriculum of settler silencing 213–15
highbrow culture 78
Hislop, J. 191–2
Hochschild, A.R. 97
homosexual activity in public toilets, observational research of 128–31
housework
    ethnicity and 43–4
    experiments and 168
    cleanfluencers 147
    gender and 168
    interviews and 97
    Oakley (Ann) study of 104–8
    observations of 122
    outsourcing of 98–9
    redundancy and 97
    social surveys 67
    statistics about 43–4
human agency 15
Humphreys, L. 128–31
hypothesis 15, 36

I

indexicality 171
Indigenous ethics 31
inductive approach 17–18
informed consent 30
insider status 118, 123
institutional racism 6
interpretivism 10, 16–17
intersectionality 19
interviews
    advantages of 120
    constructionist interviewers 104
    definition 96–7
    disadvantages of 120
    interpretivist interviewers 104
    realist interviewers 104
    reliability and validity 103–4
    sampling methods 101
    semi-structured 99
    as a social research method 120
    structured 99
    unstructured 99

J

Jackson, C. 164–5
Jacobson, L. 169
Jeon, N. 177–81
John, J. 173–4

K

Kaupapa Māori methodologies 213
Keizer, K. 182–3
Keuschnigg, M. 182–3
Khattab, N. 59
Koch, I. 45–6
Kuhn, T. 22

L

labelling 56, 88, 110–11, 196–8
Labour Force Survey (LFS) 50
Lakatos, I. 22
Lau-Layton, C. 118–19
Laurison, D. 50–54
leading questions 71
leisure gap 97
LGBTQIA+ communities 206
Likert scale 68
Lindenberg, S. 182–3
literature reviews, process of conducting 33–4
littering 182–3
Littler, J. 147
loneliness 37
low-brow culture 78
Lupton, D. 156–7, 204, 210–12

M

MacDonald, L. 213–15
macro perspective 6, 7, 15
Maguire, M. 160–1
male infertility, social construction of 92–4
Malinowski, B. 19
marketisation 148
Marx, K. 18
Marxism 21, 199
Marxist realist epistemology 18
massive open online courses (MOOCS) 173–4
Mass Observation Project 42
materialism 18
materialist theory of knowledge *see* materialism

Matsinhe, D. 220
matrix of domination 19
Maynard-Moody, S. 90–1
McArthur, D. 162–3
McDonald, S. 67
McKnight, A. 54–7
Meadows, R. 191–2
meaning 151
medicalisation 149
Meghji, A. 31
member-checking 213
meritocracy 50, 148
methodological pluralism 185, 187
methodology 11
micro perspective 6, 15, 17
middle-brow culture 78
middle-class crimes 14
middle-class professional women 115–17
Miller, T. 108–9
Mills, C. W. 209
mixed methods 9, 185
   advantages of 200
   combination of quantitative and qualitative methods 186
   data sources 186
   disadvantages of 200
   interpretivist approach 188
   positivist approach 188
   realist approach 188
   reliability and validity 186–7
   in social research 200
   theory 186–7
   triangulation by investigator 187
   types of triangulation 186
Modood, T. 59
Mohanty, C. T. 19
monism 202
Moonies, study of 189–90
Moore, S. 193–5
moral panic 9, 151, 155, 193
motherhood 18, 105–8, 115–17
multivariate analysis 48

N

narrative turn 40–1
National Pupil Database 58
neoliberalism 148
neologisms 137

non-leading questions 71
non-participant observation 9, 123
non-probability sampling 25
non-representative or purposive sampling 25
non-sociological research 13
normal science 22
normative assumptions 71

O

Oakley, A. 18, 97, 104–8
Oates, C.J. 67
objective reality 14
objectivist ontological position 15
objectivists 17
objectivity 21, 126
observational research 9
observations 17, 121
   advantages of 145
   approaches to 123
   constructionist approach 127
   disadvantages of 145
   'emic' and 'etic' perspectives 124
   forms of observational enquiry 123
   gaining access to field 125
   on housework 122
   interpretivist approach 127
   non-participant 123
   observational skills, tools and dispositions 125–6
   participant 123
   positivist approach 127
   realist approach 127
   as a social research method 145
   validity and reliability 126
Office for National Statistics (ONS) 42–4
official statistics 42–4
   about domestic violence 63–4
   advantages of 65
   constructionist sociologists and 47
   data reliability and validity 46–7
   data sufficiency 46
   disadvantages of 65
   in education 58–9
   interpretivist sociologists and 47
   of police powers to stop and search individuals 60–2
   positivist sociologists and 47
   realist sociologists and 47
   as a social research method 65
   sociologists use of 44–5

ontology 14–15
open-ended questions 70
open questions 8
open society 53
operationalise 15
Orgad, S. 115–17
othering culture 19
'othering' of British Muslims 151
outsider status 123
over-immersion 126

P

Pande. R. 136–9
paradigm shift 22
parenting in British Chinese families 8
participant observation 9, 123
patriarchy 6, 17
Pedersen, S. 156–7
Phillips, C. 60–2, 110–12
pilot survey 76, 97
plutocracy 148
podcasts 146
police
    powers to stop and search individuals 60–2
    pulled over by 90–1
policy analysis 20
Poole, E. 150–1
Popper, K. 21
population 23
positionality 38
positivism 10, 15–16, 21
positivist quantitative questions 36
positivist researchers 10, 47, 159, 187
post-humanism 202
postmodernism 20
post-qualitative inquiry 204
post-qualitative methods 202
post-structuralism 20
poverty
    media portrayal of 162–3
    in United Kingdom 76–7
Powell, B. 177–81
predictive questions 35
primary data 8
probability sampling 24–5
proletariat 18
pseudo-sociology 13

Q

Quadlin, N. 177–81
qualitative data 66
qualitative observations 9
qualitative research 6–7
    judgment criteria in 29
qualitative thematic analysis 39
quantitative data 66
quantitative research 6–7
quantitative statistical analysis 39
quare studies 205
quasi-qualitative method 8
queering methods 11, 205–6
    queer scavenger methods 206
queer theory 11
questionnaire 66, 68
quota sampling 26

R

race/racism 17, 19
    racial discrimination in America 74–5
    Gypsy Traveller communities and 110–11
    education and 59, 112–14, 172–4
    media reporting and 150–1, 154–5
random/probability sampling 24–5
realism 15, 17–19, 21
Reeves, A. 162–3
reflexivity 26, 38, 125–6
relative poverty 76
reliability 8, 28, 46–7
    documentary methods 151–2
    experimental methods 169
    interviews 103–4
    mixed methods 186–7
    observations 126
    official statistics 46–7
    social surveys 73
representativeness 23–7, 151
researcher insights 8, 34
    documentary analysis of policy 148–9
    interviews 101–3
    observations 127
    social surveys 73
    use of official statistics 45–6
    value of mixed methods 188
research methods 6
research programmes 22
research questions, formulating 34–6

right wing 152
risk discourse 164
Rosenthal, R. 169

S

samples 16
sampling 23
   convenience 25–6
   frame 24
sampling method 25
   convenience 26
   non-probability 25
   non-representative or purposive 25
   probability 24–5
   random 25
   sample size and response rate 26–7
   snowball 26
   stratified 24
saturation 27
Savage, M. 81–5, 217–18
Szabo, M. 122
Schmeid, V. 210–12
secondary data 8
second shift 97
self-administered or self-completed survey 66
self-confidence, attainment of 86–9
self-fulfilling prophecy 86–9
semi-structured interviews 17, 97–9, 118, 120, 210
settler-colonial societies 213
settler silencing 213–15
sex in public places 128–31
Shain, F. 59, 112–14
Sharma, U. 101–3
Shildrick, M. 210
Shilling, C. 203
Sikh ethics 31
Sikhism 31
Singha, L. 98–9
situational ethics 147
sleep disruption, gendered nature of 191–2
Smart, C. 203
snowball prophecy 89, 102
snowball sampling 25–6
social class
   composition of 78–85
   education and 54–58,
   occupation and 50–54
social desirability thesis 175

social facts 16
social inequality 4, 8, 10, 19, 20, 25, 41, 44, 45, 47, 49, 50–4, 54–7, 73, 77, 80, 81–5, 107, 117, 134, 140-1, 142–4, 157, 162–3, 218
social injustice 18, 19, 31
socialisation 80, 118,
social media analysis 20
social mobility 50, 54
social research 4–5
   approaches to 6–7
   background research 33–4
   choosing methods 37–8
   data gathering and analysis 39–41
   formulating research question 34–6
   operationalising 37
   practicalities of 6
   research ethics 38
   rules for writing up 40–1
social researcher, becoming 36, 38, 43, 60, 72, 100, 124–5, 171
   ethical standing 30
   feminist perspectives 10
   influencing factors 10–11
   interpretivist sociologists 10
   positivist researcher 10
   research positionality 11
social surveys 66
   advantages of 95
   analysis 72
   choosing question type 68–70
   disadvantages of 95
   face-to-face or self-completed 72
   on housework 67
   operationalising a research question 67–8
   positivist sociologists and 73
   reliability and validity 73
   sample and sample size 72
   as social research method 94
   wording of questions 70–1
sociological enquiry 6, 9, 12, 18, 206
sociological imagination 9, 193, 218–20
   see also STRIVE rules of research methods
sociological knowledge 23, 144, 205, 217, 219
sociological research 23
sociologists
   data gathering 14
   datasets and its interpretation 14
   perspective on social world 12

role in questioning and holding pseudo-sociology 13
Stacey, L. 175–6
Standard Assessment Test (SAT) 196
standard deviation 39
standpoint position 18
Stanford Prison Experiments 167
statistical artefact 47
statistical data 45
statistical discrimination 62
statistical iceberg 47
statistical software 39
Steg, L. 182–3
stigma 47
Strand, S. 58
stratified sampling 24
STRIVE rules of research methods 7–8, 12, 44, 218–20
    ethical guidelines 30, 220
    impactful 27, 219
    reflexivity 30–2, 220
    reliability 28, 219–20
    representativeness 23–7, 219
    sociological elements 12–14, 218
    theoretical perspective 14–20, 219
    validity 27–8, 219–20
structured interviews 99
subculture 128–31, 199
suicide 48–9
surveillance society 42
symbolic interactionism 17

T

tastes and cultural practices 78–80
Taylor, R. 158–9
thick description 125
Thebauld, S. 168
Thomas, G. 156–7
Towers, J. 63–4
Townsend, P. 76–7
transcription 100
transferability 29
transformative questions 35
transgender studies 205
trustworthiness 29
Tuskegee Experiment 167

U

Ubuntu ethics 30
UK Census 28
UK Household Longitudinal Study 43
unblinkered questions 71
uncanny knowledge 203
unstructured interviews 16–17, 99
Uprichard, E. 164–5

V

validity 8, 27–8, 46–7, 219–20
    documentary methods 151–2
    experimental methods 169
    interviews 103–4
    mixed methods 186–7
    observations 126
    official statistics 46–7
    social surveys 73
variables 48
Venkatesh, S. 132–5
Venn, S. 191–2
Verstehen 16–17

W

Walby, S. 63–4
Watson, A. 204
Weber, M. 17
'western' perspective 19, 32, 136
white-collar crimes 18
White normativity and White privilege 71, 118, 213
Williamson, M. 150–1
Willis, P. 199
Wolbring, T. 182–3
women's experiences of childbirth 105–8, 210–12
writing up research 126

X

xenophobia 207

Y

Youth Cohort Survey 58

Z

Zimbardo, P.G. 167
zombie categories 11

ACKNOWLEDGEMENTS

We are grateful to the following for permission to reproduce copyright material:

Texts, tables and figures

'Definition of loneliness' on p.37 from *National Measurement of Loneliness: 2018*, https://www.ons.gov.uk/peoplepopulationandcommunity/wellbeing/compendium/nationalmeasurementofloneliness/2018. Content available under the Open Government Licence v3; Table 3.1 'Number of LFS respondents in each elite occupation by origin' on p.52 and extracts on pp.52–3 from *The Class Ceiling: Why it Pays to Be Privileged* by Sam Friedman and Daniel Laurison, Policy Press, copyright © Policy Press, 2020. Reproduced by the Licensor through PLSClear; Table 3.3 on p.64 'Estimated numbers of violent crimes (violence against the person and sexual offences) by domestic, acquaintance or stranger, by sex of victim, CSEW, 2011/2, capped and uncapped' from 'The decline in the rate of domestic violence has stopped: Removing the cap on repeat victimisation reveals more violence' by Professor Sylvia Walby, Dr Jude Towers and Professor Brian Francis in *Violence & Society Research Briefing*, Nov 2014, Lancaster University. Reproduced by kind permission of the authors; Table 4.2 on p.79 'Tastes and cultural practices of classes and class fractions' from *Distinction: A Social Critique of the Judgement of Taste* by Pierre Bourdieu, translated by Richard Nice, Routledge, 2010, copyright © Pierre Bourdieu, 1984, 2010. Reproduced by arrangement with Taylor & Francis Group; Tables 4.4 'Summary of social classes' on p.83 and 4.5 'Socio-demographic correlates of seven classes' on p.84 from *Social Class in the 21st Century* by Mike Savage, Pelican, 2015. Reproduced by kind permission of the author and Penguin Books Ltd; Figure 4.1 on p.88 'Trends in self-confidence over two years when comparing students in top set and bottom set with those in middle sets' from 'The impact of tracking by attainment on pupil self-confidence over time: demonstrating the accumulative impact of self-fulfilling prophecy' by Becky Francis, Nicole Craig, Jeremy Hodgen, Becky Taylor, Antonina Tereshchenko, Paul Connolly & Louise Archer in *British Journal of Sociology of Education*, 41(5), Taylor & Francis, 03/07/2020, pp.626–42. https://www.tandfonline.com/doi/full/10.1080/01425692.2020.1763162. Reproduced with permission of Taylor & Francis; Figure 4.2 on p.91 'Officer demeanour, by race of driver, in investigatory stops' from *Pulled over: How police stops define race and citizenship* by Charles R. Epp, Steven Maynard-Moody and Donald P. Hader-Markel, University of Chicago Press, 2014, copyright © The University of Chicago, 2014; Extracts on pp.93, 94 from 'The social construction of male infertility: a qualitative questionnaire study of men with a male factor infertility diagnosis' by Esmée Hanna & Brendan Gough, *Sociology of Health & Illness*, Vol 42(3), John Wiley & Sons, 27/11/2019, pp.465–80, copyright © Foundation for the Sociology of Health & Illness, 2019. https://doi.org/10.1111/1467-9566.13038. Reproduced with permission; Extracts on pp.101, 102 from 'Reflexivity, ethnography and the professions (complementary medicine) watching you watching me watching you (and writing about both of us)' by Sarah Cant and Ursula Sharma, *The Sociological Review*, Vol 46(2), SAGE Publications, 01/05/1998, pp.244–63, copyright © SAGE Publications, 1998, https://doi.org/10.1111/1467-954X.00118; Extracts on pp.105–7 from *From Here to Maternity: Becoming a Mother* by Ann Oakley, Policy Press, 2019, copyright © Ann Oakley, 2019; An extract on p.108 from *The Sociology of Housework* by Ann Oakley, first published in 1974. This edition by Policy Press, 2008, copyright © Ann Oakley, 2018; Extracts on pp.112, 113 from *The New Folk Devils: Muslim Boys and Education in England* by Farzana Shain, Trentham Books, 2011. Reproduced with kind permission of the author; An extract on p.114 from 'Navigating the unequal education space in post-9/11 England: British Muslim girls talk about their educational aspirations and future expectations' by Farzana Shain in *Educational Philosophy and Theory*, Vol 53(3), Taylor & Francis, 23/02/2021, pp.270–87. Reproduced with permission of the author and Taylor & Francis Ltd, http://www.tandfonline.com; Extracts on pp.115–17 from *Heading Home: Motherhood, Work, and the Failed Promise of Equality* by Shani Orgad, Columbia University Press. 2019, copyright © Shani Orgad, 2019. Reproduced by permission of Columbia University Press; Figure 6.1 on p.123 'Participant Observation Continuums' from *Collecting Qualitative Data: A Field Manual for Applied Research* by Greg Guest, Emily E. Namey & Marilyn L. Mitchell, SAGE Publications Ltd, 2013, Figure 3.1, p.16, copyright © SAGE Publications Ltd, 2013. Permission conveyed through Copyright Clearance Center; Extracts on pp.128, 131 from *Tearoom Trade: Impersonal Sex in Public Places, 2nd edition* by Laud Humphreys, Routledge, 1975, copyright © R. A. Laud Humphreys, 1970, 1975. Reproduced by arrangement with Taylor & Francis Group; Extracts on pp.132–4 from *Gang Leader for a Day: A Rogue Sociologist Takes to the Streets* by Sudhir Venkatesh, Allen Lane, copyright © Sudhir Venkatesh, 2008. Reproduced by permission of Penguin Books Ltd; and Penguin Press, an imprint of Penguin Publishing Group, a division of Penguin Random House LLC. All rights reserved; Extracts on pp.136, 138, 139 from *Learning to Love: Arranged Marriages and the British Indian Diaspora* by Raksha Pande, Rutgers University Press, 2021. Reproduced with permission; Extracts on pp.142, 144 from *Hunger Pains: Life inside foodbank Britain* by Kayleigh Garthwaite, Policy Press, 2016, copyright © Policy Press, 2016; An extract on p.147 from *Ethics Guidelines and Collated Resources for Digital Research*, British Sociological Association, https://www.britsoc.co.uk/media/24309/bsa_statement_of_ethical_practice_annexe.pdf. The BSA is the British national subject association for Sociology, set up in 1961. www.britsoc.co.uk. Reproduced with permission; Table 7.1 on p.150 'Number of articles referencing Muslims and Islam 30/3/2020 to 30/04/2020', and an extract on p.151 from 'Disrupting or reconfiguring racist narratives about Muslims? The representation of British Muslims during the Covid crisis' by Elizabet Poole and Milly Williamson, *Journalism*, Vol 24(2), SAGE Publications, 02/07/2021, pp.262–79, copyright © SAGE Publications, 2021 https://doi.org/10.1177/14648849211030129; An extract on p.153 from *Discipline and Punish: The Birth of the Prison* by Michel Foucault, translated by Alan Sheridan, Penguin Classics. Originally published as *Surveiller et punir: Naissance de la prison* by Michel Foucault, 1975, copyright © Editions Gallimard, 1975, translation copyright © Alan Sheridan, 1977. Reproduced by permission of Penguin Books Ltd; Editions Gallimard; and Pantheon Books, an imprint of the Knopf Doubleday Publishing Group, a division of Penguin Random House LLC. All rights reserved; Table 7.2 on p.155 'The press coverage of "mugging" (August 1972 to August 1973)' from *Policing the Crisis: Mugging, the State and Law & Order* by Stuart Hall, Chas Critcher, Tony Jefferson John Clarke and

Brian Roberts, Red Globe Press, 2013, copyright © Stuart Hall, Chas Critcher, Tony Jefferson John Clarke and Brian Roberts. Reproduced by permission of Bloomsbury Publishing Plc.; An extract on p.156 from '"The appy for a happy pappy": expectant fatherhood and pregnancy apps' by Gareth M. Thomas, Deborah Lupton and Sarah Pedersen, *Journal of Gender Studies*, Vol 27(7), Taylor & Francis, 03/10/2018, pp.759–70. Reproduced with permission; Table 7.3 on p.163 'Descriptive statistics for words measuring stigmatising rhetoric about the poor' from 'The Rhetoric of Recessions: How British Newspapers Talk about the Poor When Unemployment Rises, 1896–2000' by Daniel McArthur and Aaron Reeves, Sociology, 56(3), SAGE Publications, 01/12/2019, pp.1005–25, Table 1, copyright © 2019, © SAGE Publications. Reproduced with permission of the publisher and author; Figures 8.1, 8.2 '(a) Instructor response' and '(b) peer response' on p.173 from 'Bias in online classes: Evidence from a field experiment' by Rachel Baker, Thomas Dee, Brent Evans and June John, *Economics of Education Review*, Vol 88, June 2022, 102259, copyright © 2022 Elsevier Ltd. All rights reserved; Extracts on p.176 from '(Bio)logics of the Family: Gender, Biological Relatedness, and Attitudes Toward Children's Gender Nonconformity in a Vignette Experiment' by Lawrence Stacey, *Sociological Forum*, Vol.37(1), John Wiley & Sons, March 2022, pp.222–45, copyright © Eastern Sociological Society, 2021. Reproduced with permission; Extracts on p.178 from 'Untangling perceptions of atypical parents' by Natasha Quadlin, Nanum Jeon, Long Doan and Brian Powell, *Journal of Marriage and Family*, Vol.84(4), John Wiley & Sons, 08/08/2022, pp.1175–95, copyright © National Council on Family Relations, 2022. Reproduced with permission; Table 8.1 on p.179–80 'Descriptive Statistics – outcome variables' in 'Untangling perceptions of atypical parents' by Natasha Quadlin, Nanum Jeon, Long Doan and Brian Powell, *Journal of Marriage and Family*, 84(4), John Wiley & Sons, August 2022, pp.1175–95, copyright © National Council on Family Relations, 2022. Source: *Constructing the Family and Higher Education Survey*, 2015. Reproduced with kind permission of the authors and publisher; Table 8.2 on p.183 'Number of cases across dormitories and experimental conditions' from 'Disorder, social capital, and norm violation: Three field experiments on the broken windows thesis' by Marc Keuschnigg and Tobias Wolbring, *Rationality and Society*, 27(1), SAGE Publications, 01/02/2015, pp.96–126, copyright © SAGE Publications, 2015; Figure 9.1 'The four shifts' and extract on p.192 from 'The fourth shift: exploring the gendered nature of sleep disruption among couples with children' by Susan Venn, Sara Arber, Robert Meadows and Jenny Hislop, *The British Journal of Sociology*, Vol 59(1), John Wiley & Sons, March 2008, pp.79–97, https://doi.org/10.1111/j.1468-4446.2007.00183.x. Open Access; Table 9.1 on p.194 'Respondents were asked the following question: "Under what circumstances do you consider yourself to be most at risk from sexual assault?"' and extract on pp.194–5 from 'Embodying Uncertainty?: Understanding Heightened Risk Perception of Drink "spiking"' by Adam Burgess, Pamela Donovan and Sarah E.H. Moore, *British Journal of Criminology*, 49(6), Oxford University Press, 29/07/2009, pp.848–64, Copyright © 2009, Oxford University Press; Extracts on pp.196, 198 from 'The role of the Boffin as Abject other in Gendered Performances of School Achievement' by Becky Francis, *The Sociological Review*, 57(4), SAGE Publications, 01/11/2009, pp.645–69, copyright © SAGE Publications, 2009; Extracts on p.204 from 'Towards more-than-human digital data studies: developing research-creation methods' by Deborah Lupton and Ash Watson, *Qualitative Research*, 21(4), SAGE Publications, 01/08/2021, pp.463–80, copyright © SAGE Publications, 2021; Extracts on p.208 from 'Becoming a skinhead: An ethnobiography of brutalized life and reflective violence' by Jérôme Beauchez, *The Sociological Review*, 69(6), SAGE Publications, 01/11/2021, pp.1179–94, copyright © SAGE Publications, 2021; Extracts on pp.211–12 from 'Splitting bodies/selves: Women's concepts of embodiment at the moment of birth' by Deborah Lupton and Virginia Schmeid, *Sociology of Health & Illness*, 35(6), John Wiley & Sons, 24/10/2012, pp.828–41, copyright © The Authors and Foundation for the Sociology of Health & Illness/John Wiley & Sons Ltd, 2012; An extract on p.214 from '"The same as everyone else": How academically successful indigenous secondary school students respond to a hidden curriculum of settler silencing' by Liana MacDonald, *Whiteness and Education*, Vol 4(1), Taylor & Francis, 02/01/2019, pp.38–82. https://doi.org/10.1080/23793406.2019.1626758. Reproduced with permission of Taylor & Francis; Extracts on p.217 from 'The Coming Crisis of Empirical Sociology' by Mike Savage and Roger Burrows, *Sociology*, 41(5), SAGE Publications, 01/10/2007, pp. 885–99, copyright © SAGE Publications, 2007; and an extract on p.220 from 'Quest for Methodological Alternatives' by David Mario Matsinhe, *Current Sociology*, 55(6), SAGE Publications, 01/11/2007, pp.836–56, copyright © SAGE Publications, 2007.

Images

Cover: Laura Buckley, *Fata Morgana*, 2012 © Laura Buckley, photo: emmak95/Stockimo/Alamy Stock Photo; p.4 creativeneko/Shutterstock; p.7 Maksym Drozd/Shutterstock; p.11 Magenta10/Shutterstock; p.13 Skylines/Shutterstock; p.16 Natata/Shutterstock; p.17 Natata/Shutterstock; p.18 Everett Collection/Shutterstock; p.18 B. Gomer / Stringer / Getty Images; p.19 IanDagnall Computing / Alamy Stock Photo; p.19 howcolour/Shutterstock; p.24 petrroudny43/Shutterstock; p.25 Bakhtiar Zein/Shutterstock; p.25 Bakhtiar Zein/Shutterstock; p.26 Bakhtiar Zein/Shutterstock; p.26 Bakhtiar Zein/Shutterstock; p.34 acreative/Shutterstock; p.38 tete_escape/Shutterstock; p.42 mundissima/Shutterstock; p.46 cktravels.com/Shutterstock; p.51 Andrey_Popov/Shutterstock; p.57 jamesteohart/Shutterstock; p.59 Anel Alijagic/Shutterstock; p.62 Loch Earn/Shutterstock; p.66 Irina Strelnikova/Shutterstock; p.68 dotshock/Shutterstock; p.69 Yeexin Richelle/Shutterstock; p.74 Public domain; p.83 © Andy Watt/The Independent; p.96 eggeegg/Shutterstock; p.98 Pixel-Shot/Shutterstock; p.100 fizkes/Shutterstock; p.109 Destiny13/Shutterstock; p.121 BadBrother/Shutterstock; p.122 Monkey Business Images/Shutterstock; p.126 fizkes/Shutterstock; p.135 The Color Archives / Alamy Stock Photo; p.141 Africa Studio/Shutterstock; p.143 HASPhotos/Shutterstock; p.146 BRO.vector/Shutterstock; p.147 Nopparat Khokthong/Shutterstock; p.149 GaudiLab/Shutterstock; p.152 Lenscap Photography/Shutterstock; p.157 Friends Stock/Shutterstock; p.159 Feng Yu/Shutterstock; p.161 Daisy Daisy/Shutterstock; p.166 Andrew Krasovitckii/Shutterstock; p.167 Skimage / Alamy Stock Photo; p.172 Kitreel/Shutterstock; p.174 Rawpixel.com/Shutterstock; p.176 galitsin/Shutterstock; p.178 wavebreakmedia/Shutterstock; p.181 Littlekidmoment/Shutterstock; p.185 picotan/Shutterstock; p.187 alphaspirit.it/Shutterstock; p.191 KieferPix/Shutterstock; p.193 PeopleImages.com - Yuri A/Shutterstock; p.197 Monkey Business Images/Shutterstock; p.199 Victor de Schwanberg / Alamy Stock Photo; p.202 MedRocky/Shutterstock; p.203 Renata Apanaviciene/Shutterstock; p.204 Andrei Krushko/Shutterstock; p.206 Pixel-Shot/Shutterstock; p.208 Suzanne Tucker/Shutterstock; p.211 Demkat/Shutterstock; p.220 Badon Hill Studio/Shutterstock